AI EVERYDAY

Transforming Lives with Smart Technology

by
William Scott

AI EVERYDAY

Transforming Lives with Smart Technology

CONTENTS

INTRODUCTION

As we march into the 21st century, the transformative power of Artificial Intelligence (AI) permeates nearly every aspect of our daily lives. From the moment we wake up to an AI-powered alarm clock to the smart assistant that checks the weather and clears our schedules, AI is becoming an indispensable part of our routines. Yet, despite its ubiquity, many still view AI as a futuristic technology confined to the realms of science fiction and advanced research labs. This book aims to dismantle those misconceptions by elucidating the practical, real-world applications of AI and how they can enhance everyday life.

While the idea of AI might conjure images of self-aware robots or complex algorithms, the essence of AI lies in its ability to mimic human intelligence. This superhuman capability manifests through learning, reasoning, problem-solving, and adapting. What sets AI apart is its ability to perform these tasks at unprecedented speed and efficiency, often surpassing human capabilities. But this isn't a story of humans versus machines; it's a narrative of how technology can work hand-in-hand with us to craft a more convenient, safer, and enriching life.

Imagine waking up in a home that 'knows' you. From adjusting the room temperature to preparing your morning coffee just the way you like it, smart home systems powered by AI make this possible. These innovations aren't mere conveniences; they represent a significant leap toward a lifestyle that prioritizes efficiency and personalization.

In healthcare, AI is not only augmenting diagnostics but also paving the way for personalized medicine. Procedures that previously required hours of a doctor's attention can now be completed faster and more accurately, thanks to AI-driven tools. Health monitoring devices powered by AI alert us to potential illnesses before they become critical, essentially acting as vigilant guardians of our well-being.

Education, too, stands at the cusp of a revolution, led by AI. Adaptive learning technologies provide personalized educational experiences, tailoring lessons to individual learning styles and paces. Virtual classrooms and AI tutors democratize education, making quality learning accessible to everyone, irrespective of geographical barriers or socioeconomic status.

Possibly, one of the most impressive contributions of AI is in the realm of personal finance. Automated budgeting tools help manage finances more efficiently, while AI-driven investment platforms offer tailored advice, previously reserved for high-net-worth individuals serviced by exclusive financial advisors. Furthermore, advanced fraud detection systems help safeguard our hard-earned money, bringing peace of mind.

Entertainment has never been more personalized. AI algorithms that suggest movies, music, or even books, know our preferences better than we do. Video games powered by AI create dynamic experiences that evolve based on how we play. Virtual reality, enhanced by AI, offers immersive experiences that blur the line between the real and virtual worlds.

In the sphere of communication, AI has transformed how we interact and exchange information. Innovations in natural language processing make our conversations with customer service bots nearly indistinguishable from those with human representatives. Smart messaging apps predict and suggest responses, making our daily communications more fluid and less time-consuming.

When it comes to transportation, AI is at the forefront of developing autonomous vehicles. These self-driving cars promise to reduce accidents, alleviate traffic congestion, and provide mobility options for those unable to drive. Intelligent traffic management systems are already optimizing traffic flow in cities worldwide, and AI-powered ride-sharing services are offering more efficient and eco-friendly transportation solutions.

Shopping has been revolutionized by AI in ways that make our experiences more personalized and efficient. From recommending products based on previous purchases to optimizing supply chains and even managing smart warehouses, AI is reshaping the retail landscape.

In the workplace, productivity tools powered by AI are becoming indispensable. Intelligent task management systems organize our priorities, and AI in project management ensures that we meet deadlines more efficiently. Virtual assistants help manage our schedules, respond to emails, and handle various tasks, freeing us to focus on what truly matters.

On social media platforms, AI helps filter content, providing us with personalized experiences tailored to our interests. It also aids in analyzing vast amounts of data, offering insights that can optimize personal and business strategies. Additionally, bots and automation streamline operations, enhancing user interactions.

Even in real estate, AI is making significant strides. Smart property management systems ensure homes are more energy-efficient and secure. AI-assisted home buying and selling platforms simplify the process, providing virtual tours that afford potential buyers the luxury of viewing multiple properties without leaving their homes.

Travel and tourism are sectors ripe for AI intervention. Personalized travel planning systems curate itineraries based on preferences and past behaviors. AI in customer service enhances traveler experiences, and smart travel apps make traveling more convenient and enjoyable.

Moreover, fitness and wellness have benefited tremendously from AI. Personalized workout plans ensure that we achieve our health goals more effectively, and AI in mental health provides resources and support that cater to our emotional well-being. Smart wearables track our physical activity, offering insights that help us lead healthier lives.

The agriculture industry sees AI's potential in precision farming, where monitoring and automating crop management ensure higher yields and better resource utilization. From choosing the right time to plant to predicting weather patterns, AI's role in agriculture is becoming indispensable.

Environmental sustainability is another critical area where AI's influence is growing. Smart energy management systems optimize the use of resources, and AI-driven waste management solutions ensure that waste is minimized. Environmental monitoring powered by AI helps in maintaining ecological balance and addressing climate change challenges.

Governments are also harnessing AI to improve public services. Smart cities incorporate AI to enhance public safety, manage resources, and provide digital services that make life more convenient for citizens. Such innovations pave the way for a more efficient and responsive governance model.

In the legal field, AI assists in legal research, providing quick and accurate information that would take human researchers hours or days to collate. Intelligent contract analysis tools ensure that legal documents are meticulous, and predictive analytics help in formulating strategies and making informed decisions.

Marketing has been transformed by AI, enabling more targeted advertising that reaches the right audience at the right time. Customer insights derived from AI analytics help businesses understand their market better, while content generation tools streamline the creation of engaging and relevant material.

Human resources departments are increasingly relying on AI for intelligent recruitment processes, ensuring the best talent is identified and hired. AI in employee training tailors learning experiences based on individual needs, and performance monitoring tools offer regular, unbiased feedback, fostering a more productive work environment.

Manufacturing benefits tremendously from AI, with predictive maintenance ensuring that machinery operates seamlessly. Robotics streamline processes, while quality control systems managed by AI guarantee that products meet the highest standards.

In logistics and supply chain management, AI optimizes inventory management, making sure that products are always in stock and delivered on time. Route optimization powered by AI reduces delivery times, and demand forecasting helps businesses plan more effectively.

The energy and utilities sector witnesses AI's potential in smart grids, which efficiently manage energy distribution. AI in energy forecasting predicts consumption patterns, ensuring better resource allocation and minimizing waste.

While the advantages of AI are numerous

CHAPTER 1:
UNDERSTANDING ARTIFICIAL INTELLIGENCE

Understanding Artificial Intelligence (AI) begins with recognizing its transformative potential to revolutionize various aspects of our daily lives. At its core, AI encompasses a broad range of technologies that enable machines to perform tasks requiring human intelligence, such as learning, reasoning, problem-solving, and decision-making. By leveraging vast amounts of data and sophisticated algorithms, AI systems can analyze patterns, predict outcomes, and even adapt to new information autonomously. As we delve into the complexities of AI, we'll explore its underlying principles, key concepts, and terminologies, setting a foundation for appreciating how this technology can simplify, enhance, and enrich our everyday experiences. Whether it's through smart assistants, personalized healthcare, or adaptive educational tools, AI is poised to become an integral part of our lives, offering unprecedented opportunities for efficiency and innovation in all sectors.

What is AI?

Artificial Intelligence, often abbreviated as AI, is a term widely used yet not universally understood. Broadly speaking, AI refers to the simulation of human intelligence in machines that are programmed to think and learn. These machines can perform a variety of tasks that typically require human intelligence, such as recognizing speech, making decisions, solving problems, understanding natural language, and even emulating human emotions. Unlike traditional software, which follows a predetermined set of instructions, AI systems can adapt and improve from experience, virtually learning on the job.

The idea of creating intelligent machines has been around for centuries, dating back to ancient myths and legends about mechanical beings. However, the modern era of AI began in the mid-20th century, marked by a surge of research and technological advancements. Early pioneers like Alan Turing and John McCarthy laid the groundwork, envisioning computers capable of human-like thought processes. Turing proposed the famous "Turing Test" to determine a machine's ability to exhibit intelligent behavior indistinguishable from a human's. McCarthy, on the other hand, organized the Dartmouth Conference in 1956, a pivotal event that is often credited with launching the field of AI.

The core of AI encompasses several key principles. At its foundation is the concept of **machine learning**, a technique that enables machines to learn from data. This approach is in stark contrast to traditional programming where explicit instructions are coded for the machine to follow. Machine learning algorithms, often underpinned by statistical models, iteratively learn patterns in data and use these patterns to make predictions or decisions. There are various types of machine learning, including supervised learning, unsupervised learning, and reinforcement learning, each with its own set of methodologies and applications.

Another critical component of AI is *deep learning*, a subset of machine learning that uses neural networks with many layers (hence "deep") to analyze data. Deep learning has driven many of the recent breakthroughs in AI, enabling technologies such as image and speech recognition, autonomous driving, and advanced natural language processing. These neural networks are inspired by the human brain's structure and function, containing nodes (analogous to neurons) that process information in layers. The depth of these networks allows them to capture complex patterns and representations, making them particularly powerful for tasks like recognizing faces or understanding context in language.

Natural Language Processing (NLP) is another area where AI has shown extraordinary capabilities. NLP involves enabling machines to understand, interpret, and generate human language. This technology powers applications like chatbots, translation services, and virtual assistants, bridging the gap between human communication and machine understanding. Techniques like sentiment analysis, named entity recognition, and machine translation fall under the realm of NLP, making it one of AI's most visible and interactive facets.

AI doesn't just stop at learning and language. It also includes **robotics**, where physical machines or robots are designed to perform human-like tasks. Robotics combined with AI leads to machines that can navigate and manipulate their environment autonomously. Examples include robotic vacuum cleaners, drones, and industrial robots that can work alongside humans in factories. The integration of AI into robotics allows these systems to learn and adapt to new tasks, making them more versatile and efficient.

One must also consider the role of AI in *decision-making systems*. These systems utilize data, analytics, and rules to make informed decisions. For example, AI algorithms are used in finance to detect fraudulent transactions, in healthcare to assist in diagnosis and treatment plans, and in marketing to identify target audiences and optimize campaigns. These decision-making systems rely on large datasets, advanced analytics, and often, real-time processing to deliver accurate and timely decisions.

To understand AI, it's essential to recognize that it's not a monolithic entity but a collection of technologies and techniques. These include machine vision, which allows machines to interpret and analyze visual information from the world, and expert systems, which mimic the decision-making abilities of a human expert. The true power of AI lies in its ability to combine these different technologies to create systems that are greater than the sum of their parts.

As AI continues to evolve, its implications for everyday life grow more profound. In essence, AI is about creating intelligent agents that can assist, augment, or automate tasks in ways that improve efficiency and effectiveness. It's becoming increasingly embedded in the fabric of our daily routines, often in invisible yet impactful ways. From personalized recommendations on streaming services to intelligent route planning in navigation apps, AI's influence is pervasive and growing.

Understanding AI requires not just a grasp of its technical underpinnings but also an appreciation for its potential and limitations. AI excels in tasks where large amounts of data are available and where patterns can be discerned. However, it still struggles with understanding the nuances of human emotions, creativity, and contextual knowledge. It's a powerful tool but not a panacea for all problems. As we continue to integrate AI into our lives, it will be crucial to maintain a balanced view, recognizing both its capabilities and its constraints.

In summary, AI represents a profound shift in the way we interact with technology and is poised to revolutionize many aspects of our daily lives. By leveraging techniques like machine learning and deep learning, natural language processing, and robotics, AI systems can learn, adapt, and perform tasks that were once thought to be solely within the human domain. As we explore further chapters, you'll discover the myriad ways AI is already transforming home automation, healthcare, education, finance, and beyond, offering a glimpse into a future where intelligent machines work alongside us to enhance our lives.

Key Concepts and Terminologies

Delving into the intricacies of Artificial Intelligence (AI) requires a solid grasp of key concepts and terminologies that underpin this transformative technology. To comprehend how AI can augment daily life, it's crucial to build a foundational understanding of these core elements. This section will serve as your guide to navigate the often complex jargon and nuanced ideas in AI.

Artificial Intelligence (AI) at its core refers to the simulation of human intelligence processes by machines, especially computer systems. These processes include learning (the acquisition of information and rules for using the information), reasoning (using rules to reach approximate or definite conclusions), and self-correction. AI can be categorized into two types: Narrow AI, which is designed and trained for a specific task, like Siri or Alexa, and General AI, which, in theory, would have the ability to perform any intellectual task that a human can do.

One crucial terminology in AI is **Machine Learning (ML)**. ML, a subset of AI, refers to the method of teaching machines to make decisions from data. Instead of being explicitly programmed to perform a task, these machines are trained using large amounts of data to recognize patterns and make decisions. For instance, recommendation systems on streaming services, which suggest movies based on your viewing history, are a practical application of ML.

Another essential concept is **Deep Learning**. This subset of ML leverages neural networks with many layers (hence, "deep") to analyze various factors of data. Deep Learning excels at processing vast amounts of data and can yield highly accurate results. Its use is prominent in applications like voice and image recognition, where it's the backbone of algorithms enabling your smartphone's ability to identify faces or understand spoken commands.

Next, let's explore **Neural Networks**. Inspired by the human brain's structure, neural networks are composed of layers of nodes, or "neurons," that filter and transmit information. Each layer processes the input data slightly differently, honing the system's ability to recognize complex patterns. A popular architecture in Neural Networks is the Convolutional Neural Network (CNN), which is particularly effective in image and video recognition tasks.

One must also understand the concept of **Natural Language Processing (NLP)**. NLP is a field of AI that focuses on the interaction between computers and humans through natural language. It involves enabling machines to read, understand, and generate human languages. Applications of NLP include chatbots, translation services, and sentiment analysis tools, which can gauge customer opinions based on their social media posts.

In the realm of AI, **Algorithms** are fundamental building blocks. An algorithm is a set of rules or instructions given to an AI system to help it learn how to complete tasks on its own. Algorithms are used within every facet of AI, from the basic filtering capabilities in your email inbox that separate spam from important messages to complex predictive models that forecast the stock market.

Another significant term is **Data Mining**. It refers to the process of discovering patterns and knowledge from large amounts of data. The data sources can include databases, data warehouses, the internet, and other data repositories. Data mining is critical for AI as it provides the raw materials (data) that are necessary to train machine learning models and improve their predictive capabilities.

Supervised Learning and **Unsupervised Learning** are types of machine learning paradigms. In supervised learning, the model is trained on a labeled dataset, meaning that each training example is paired with an output label. It's commonly used for tasks like classification and regression. In contrast, unsupervised learning works with unlabeled data and tries to find hidden patterns or intrinsic structures in the input data. Clustering and association are typical tasks managed by unsupervised learning.

The concept of **Reinforcement Learning** (RL) is another area worth understanding. RL is a type of Machine Learning where an agent learns to make decisions by performing actions and receiving rewards or penalties. It's akin to training a dog with treats; the dog (agent) learns which behaviors elicit rewards and which don't. RL is behind many autonomous systems, including self-driving cars and advanced robotics.

An important term we encounter often is **Robotics**. Robotics combines AI and mechanical engineering to design and create machines capable of performing tasks that typically require human intelligence. These tasks can range from simple manufacturing processes to complex surgical procedures. The symbiosis of AI and robotics allows for the automation and optimization of various functions in numerous industries.

Understanding AI also requires familiarity with **Big Data**. Big Data refers to extremely large datasets that may be analyzed computationally to reveal patterns, trends, and associations, especially relating to human behavior and interactions. AI thrives on data - the more data an AI system has, the better it can learn and refine its processes. This synergy between Big Data and AI unveils opportunities for advancements in predictive analytics, market research, and trend forecasting.

Ethics in AI is another critical concept. As AI becomes increasingly pervasive, the ethical implications of its use cannot be overlooked. Ethical AI involves developing systems that are fair, transparent, and accountable. Issues like data privacy, algorithmic bias, and the potential for AI to replace human jobs are central to discussions about AI ethics, ensuring that technology serves humanity positively without adverse side effects.

With the rise of AI, the term **Automation** has also become more prevalent. Automation involves the use of technology to perform tasks without human intervention. In many cases, AI-driven automation can dramatically enhance efficiency and accuracy, whether it's through robotic process automation (RPA) in businesses or autonomous systems in manufacturing. By automating repetitive and mundane tasks, AI allows humans to focus on more complex and creative endeavors.

Edge AI is another emerging concept. It involves running AI algorithms locally on a hardware device rather than relying on cloud-based computing resources. This approach allows for faster data processing and better privacy protections since sensitive information does not need to be sent to the cloud. Edge AI is gaining traction in applications requiring real-time processing, like autonomous vehicles and smart IoT devices.

Transfer Learning is an exciting concept and a subset of machine learning. It focuses on storing knowledge gained while solving one problem and applying it to a different but related problem. For instance, knowledge gained while learning to recognize cars could be applied when trying to recognize trucks. This ability to transfer knowledge accelerates the training process and enhances the performance of AI systems.

Finally, the term **Artificial General Intelligence (AGI)** captures the aspirational goal of AI research. AGI would possess the ability to understand, learn, and apply knowledge across a wide range of tasks, much like a human being. While current AI systems are highly specialized and can't perform beyond their specific domain (Narrow AI), AGI

Chapter 2:
AI in Home Automation

As we transition seamlessly from understanding the basic concepts of AI, it's fascinating to see how these technologies are transforming everyday life, particularly in the home. AI in home automation brings a blend of convenience, efficiency, and security that's reshaping how we interact with our living spaces. Imagine waking up to a home that knows your morning routine—smart assistants turning on lights gradually, adjusting the thermostat to your preferred temperature, and even preparing your coffee. Intelligent lighting systems not only enhance ambiance but also optimize energy usage, while smart climate control adapts to your lifestyle in real-time. In terms of security, AI-enabled home security systems provide robust surveillance, recognizing patterns and detecting unusual activities, giving you peace of mind. These technological advancements don't just make life easier; they contribute to creating a safer, more personalized environment that adapts to your needs, illustrating the profound impact AI can have on our daily lives.

Smart Assistants

The advent of smart assistants has revolutionized the way we interact with technology in our homes. These AI-driven entities have become integral in simplifying daily tasks, providing information, and even entertaining us. Whether it's setting a timer for cooking, controlling smart home devices, or simply answering queries, smart assistants stand as silent sentinels ready to serve at a moment's notice.

One of the most notable features of smart assistants is their ability to understand and process natural language. This capability allows users to interact with them through simple voice commands, bypassing the need for more complicated interfaces. This natural language processing (NLP) technology not only makes smart assistants user-friendly but also accessible to individuals across a wide age range and varying levels of tech-savviness.

Devices like Amazon Alexa, Google Assistant, and Apple's Siri exemplify this technology's potential. These assistants are embedded in a variety of devices, from smart speakers to smartphones, and even in some household appliances. By doing so, they create a seamlessly connected ecosystem where users can command, control, and converse without ever lifting a finger.

Smart assistants aren't just about convenience; they also bring efficiency to household management. Imagine walking into your home after a long day and simply asking your assistant to adjust the lighting, play your favorite music, or even start preheating the oven. This isn't a futuristic dream—it's today's reality, made possible through AI-driven integrations.

A remarkable advantage of smart assistants lies in their adaptability and learning capabilities. Over time, these systems analyze user behavior and preferences to provide more personalized responses and suggestions. For instance, if you often ask for the weather at a particular time each morning, your smart assistant can preemptively offer that information, saving you the effort of asking.

This level of personalization extends to other areas of home management as well. For example, you can create routines that trigger multiple actions with a single command. A "Good Night" routine could lock doors, turn off lights, and set your alarm system, creating not just convenience but also enhancing household security.

Furthermore, smart assistants are continuously evolving with the integration of third-party applications and services. From ordering groceries to booking a ride, the list of tasks these assistants can handle is ever-growing. Integration with various smart home devices such as thermostats, security cameras, and even kitchen appliances, amplifies their utility.

This interconnectedness fosters a more cohesive home automation experience. The notion of a 'smart home' becomes not just about individual smart devices but a symbiotic network of devices working in harmony, orchestrated by your assistant. This leads to improved energy efficiency, better security, and an overall enhanced quality of life.

There's also the crucial aspect of accessibility. Smart assistants have provisions to aid individuals with disabilities or those who may find traditional interfaces challenging. Voice commands can take the place of physical controls, offering an inclusive way to navigate through daily activities. For the visually impaired, the ability to ask for information and receive auditory responses is invaluable.

In terms of security and privacy, concerns naturally arise. Given that these assistants continually listen for their wake words, users may worry about the possibility of unauthorized data collection. Companies like Amazon, Google, and Apple are keenly aware of these concerns and have implemented stringent privacy controls, such as muting features and data encryption, to protect user information.

The growing use of smart assistants is also paving the way for more intelligent interactions. For instance, with advancements in AI, the ability to carry on more natural and contextually aware conversations is becoming possible. This evolution is shifting the role of smart assistants from mere tools to more interactive and 'human-like' companions.

Moreover, the integration of machine learning algorithms allows these assistants to better understand context and nuance, leading to more relevant and accurate responses. Imagine asking your assistant about traffic conditions and getting not just the congestion report but also recommendations for alternative routes based on your usual commuting patterns.

Another essential facet is the integration of multiple languages and dialects, which enhances accessibility for non-English speakers and helps bridge language barriers within multilingual households. This makes smart assistants not just a tool for the tech-savvy but a helpful companion for a broader demographic.

One exciting development is the incorporation of AI in emotional recognition, which could enable smart assistants to detect user emotions based on voice tone and deliver more empathetic interactions. While in its nascent stages, emotional AI holds the promise of making these assistants even more intuitive and responsive to our needs.

Looking towards the future, the potential of smart assistants seems almost limitless. As AI technology continues to advance, these assistants will likely become even more embedded in our daily lives, offering unprecedented levels of convenience and efficiency. Innovations in quantum computing, for instance, could further enhance their processing capabilities, making them faster and more reliable.

Envision a future where smart assistants not only perform tasks but anticipate them, taking on a proactive role in our daily routines. They could analyze your calendar, assess your energy levels, and suggest breaks or productivity hacks tailored to your work habits. The impact on personal productivity and well-being could be profound.

In conclusion, smart assistants are more than just gadgets; they represent a fundamental shift in how we interact with technology in our homes. By leveraging AI's full potential, they've transformed from simple voice-activated tools to sophisticated, intuitive companions that enrich our daily lives. As technology continues to progress, the capabilities of smart assistants will expand, paving the way for even more innovative applications that enhance our quality of life in ways we have yet to fathom.

The journey of smart assistants within the realm of home automation is just beginning. With continued technological advancements, their integration will become more seamless, their functionalities more robust, and their impact more significant. They embody the profound promise of AI to not only simplify our lives but to elevate our everyday experiences to new heights of comfort and convenience.

Intelligent Lighting and Climate Control

Imagine walking into a room where the lights automatically adjust to your preferred brightness and color, or where the temperature adapts to your ideal comfort level without you having to lift a finger. This isn't science fiction; it's the power of AI in intelligent lighting and climate control. By integrating AI with IoT (Internet of Things) devices, homes are evolving into adaptive environments that can provide precise and personalized comfort.

Intelligent lighting systems go beyond simple on-off switches. They utilize sensors, AI algorithms, and user data to create a more nuanced lighting experience. An AI-driven system can learn your daily routines and preferences. For instance, it might know that you prefer dimmer lights in the morning and brighter, cooler tones in the evening. Over time, these systems can become finely tuned, adjusting the lighting automatically to improve your productivity, enhance relaxation, or even better display art and other intricate details in your home.

Consider the ambience of a dinner party. With intelligent lighting, the system can set the mood by adjusting the lighting levels and colors across different areas of your home. The dining room could have warm, inviting lights, while the living room might have subtler, softer tones. And it all happens automatically, curated by an AI that understands the event based on your calendar, routine, or even voice commands.

AI-driven climate control systems also offer transformative benefits. Traditional thermostats allow you to set temperature preferences, but they're often not responsive to real-time changes. Modern AI-enhanced systems, on the other hand, monitor your habits and environmental conditions. They can predict when you're likely to come home and adjust the climate accordingly, ensuring your house is always at the perfect temperature upon arrival.

These systems work by gathering data from various sources like weather forecasts, energy rates, and even the physical well-being of individuals in the household. If you prefer a cooler indoor climate when sleeping, the system learns this and adjusts itself. Additionally, smart climate control can redistribute heat in more energy-efficient ways, contributing to lower energy bills and a reduced carbon footprint.

The interaction between intelligent lighting and climate control is where the magic truly happens. The interplay of light and temperature significantly influences our comfort and mood. An AI that understands this can create environments that foster both physical well-being and mental relaxation. For example, during winter, a system might combine warmer lighting with increased heating, creating a cozy atmosphere that encourages relaxation and productivity.

What separates ordinary automation from intelligent systems is the ability to learn and adapt. AI algorithms excel at analyzes patterns, learning preferences, and making informed decisions. For instance, machine learning models can take into account seasonal changes, daily routines, and even specific events to fine-tune settings. Integration with other smart home components, like blinds or fans, enhances this adaptability, making the environment even more responsive to individual needs.

Privacy and data security are critical considerations in these intelligent systems. The aggregation and analysis of personal data bring up valid concerns. It's essential to choose systems with robust privacy policies and strong security measures to protect this sensitive information. Manufacturers are increasingly aware of these concerns and often build in end-to-end encryption and anonymize user data to ensure peace of mind.

The broader impact of intelligent lighting and climate control is profound. On a societal level, these technologies contribute to significant energy savings. By optimizing power consumption based on actual usage and environmental conditions, they reduce wastage and lower utility bills. On an environmental level, they help reduce carbon emissions by fine-tuning energy usage, potentially leading to a more sustainable future.

Furthermore, intelligent lighting systems can have health benefits. Proper lighting is crucial for maintaining circadian rhythms, which affect sleep quality and overall well-being. AI can orient lighting systems to mimic natural daylight patterns, helping to regulate sleep cycles and improving mental health. Likewise, optimal climate control minimizes the risk of issues like mold growth and poor air quality, creating healthier living spaces.

For those interested in implementing such systems, the market offers various options, from DIY solutions to professionally installed systems. DIY solutions like smart bulbs and thermostats are more accessible and generally require less investment. However, they might lack the sophistication and integration offered by more comprehensive, professionally installed systems. Assessing your needs and budget can help determine the best approach for your specific situation.

Looking forward, it's easy to imagine a future where homes are fully autonomous, functioning almost like living entities that care for their inhabitants. AI-powered systems could predict your needs even before you realize them, offering an unprecedented level of comfort and convenience. As these technologies evolve, they'll likely become more intuitive and integrated, redefining modern living standards.

In summary, intelligent lighting and climate control exemplify how AI is revolutionizing everyday life. These systems offer a harmonious blend of convenience, efficiency, and personalization, radically enhancing our living environments. As you consider adopting these technologies, remember the promise they hold. They don't just make our homes smarter; they make our lives richer, healthier, and more comfortable.

•

Home Security Systems

The rise of Artificial Intelligence (AI) has revolutionized many aspects of our daily lives, and home security systems are no exception. Today, AI-driven home security solutions offer a level of sophistication and convenience that was unimaginable just a few years ago. These systems not only provide robust security but also integrate seamlessly with other smart home devices, creating a cohesive ecosystem that enhances the safety and comfort of your living space.

Modern home security systems leverage AI to provide real-time monitoring and instant alerts. Traditional alarm systems often relied on a single mode of detection, like motion sensors or door/window contacts. In contrast, AI-enhanced systems incorporate multiple layers of protection, including facial recognition, behavioral analytics, and advanced anomaly detection. These features enable the system to differentiate between a genuine threat and a false alarm, reducing unnecessary disturbances.

Facial recognition technology, a compelling aspect of AI in home security, has drastically improved over the years. By using deep learning algorithms, these systems can identify familiar faces and distinguish them from potential intruders. This capability allows for personalized settings; for example, your system might disarm itself when it recognizes you and your family members, while staying alert for unknown faces.

Behavioral analytics is another critical component of AI-enhanced security systems. By continuously learning the regular patterns of activity in your home, the system can detect anomalies that may indicate a security breach. For instance, if the system knows you typically enter your home through the front door around 6 PM, it will flag any deviations from this pattern, such as someone attempting to enter through a back window at midnight.

AI's ability to analyze and respond to security footage in real-time has transformed the role of surveillance cameras. Traditional cameras only recorded footage, requiring homeowners to review hours of video to find relevant events. Modern AI cameras, however, use smart algorithms to identify and tag significant activities, alerting you immediately if they detect suspicious behavior. This feature is particularly useful for monitoring large properties or homes with multiple entry points.

Moreover, AI-powered home security systems often come with advanced threat analysis features. These systems can assess the severity of a detected threat and determine the best course of action. In some cases, they can even alert emergency services automatically, ensuring help arrives promptly. This proactive approach not only strengthens security but also provides peace of mind.

Integration with other smart home devices is another significant advantage of AI-based home security systems. For example, if your security system detects smoke or a fire, it can activate smart lighting to illuminate escape routes and unlock smart locks to facilitate a quick exit. Similarly, if your security system notices unusual activity around your home, it might turn on exterior lights or trigger alarms to deter potential intruders.

The interconnectivity of these devices also means that you can manage and monitor your home security system remotely. Using a central hub or a smartphone app, you can arm or disarm your system, lock or unlock doors, and view live camera feeds no matter where you are. This level of control and flexibility is particularly beneficial for frequent travelers or those with multiple properties.

All these enhancements come with heightened considerations for privacy and data security. As home security systems collect and analyze vast amounts of personal data, it's crucial to ensure that these systems are protected against cyber threats. Leading providers implement robust encryption methodologies and secure cloud storage solutions to safeguard your data. Users should also follow best practices, like regularly updating their systems and using strong, unique passwords, to further enhance their digital security.

The evolution of AI in home security has also democratized access to high-end security features. While top-tier systems might still carry a premium price tag, many affordable options are available that offer comprehensive security functionalities. The competition in the market has driven innovation, resulting in a wide range of products catering to different needs and budgets.

The future of AI in home security looks incredibly promising. Predictive analytics, for instance, could further enhance threat detection by forecasting potential incidents before they occur. Machine learning algorithms will continue to improve, offering even greater accuracy and reliability. Additionally, ongoing advancements in sensor technology and connectivity will expand the capabilities of these systems, making them more intuitive and user-friendly.

In essence, AI-driven home security systems represent a harmonious blend of technology and practicality. They provide robust, intelligent, and seamless security solutions that adapt to your specific needs and preferences. Whether you're at home or away, these systems ensure that your sanctuary remains protected, giving you the peace of mind to focus on what truly matters in your life.

The integration of AI into home security systems is not just about enhancing safety but also about enriching the overall living experience. As these technologies continue to evolve, they will undoubtedly play a crucial role in shaping the smart homes of the future. Embracing AI-driven security solutions is a step towards a safer, smarter, and more connected world.

CHAPTER 3:
AI IN HEALTHCARE

The transformative power of AI in healthcare is already redefining the landscape of medical practice, offering personalized medicine, advanced diagnostics, and continuous health monitoring. Imagine receiving treatments tailored specifically to your genetic profile, or diagnostic algorithms that pick up on subtleties even the most trained eyes might miss. AI-powered health monitoring devices seamlessly integrate into daily routines, tracking vital signs and providing real-time feedback, allowing for proactive health management. This synergy of technology and healthcare not only improves the accuracy of diagnoses and efficiency of treatments but also democratizes access to high-quality medical care, making it more inclusive and comprehensive. The promise of AI in healthcare is not just futuristic but is very much a present reality, fostering a more responsive, precise, and patient-centric medical ecosystem.

Personalized Medicine

Imagine a world where your healthcare is as unique as your fingerprint. This is no longer a far-fetched dream but a rapidly evolving reality, thanks to personalized medicine, a transformative application of artificial intelligence (AI) in healthcare. Personalized medicine leverages AI to tailor medical treatments and preventive measures to individual patients, considering their genetic makeup, lifestyle, and specific health conditions. Unlike traditional one-size-fits-all approaches, personalized medicine promises more effective and targeted treatments, reducing the risk of adverse reactions and improving patient outcomes.

Central to personalized medicine is the concept of precision. AI systems can analyze vast amounts of data to identify patterns that are invisible to the human eye. These patterns can include genetic mutations that predispose individuals to certain diseases, biomarkers indicating susceptibility to specific treatments, and lifestyle factors that influence health outcomes. By integrating this information, AI enables healthcare providers to develop customized treatment plans that address the unique needs of each patient.

One of the most exciting applications of AI in personalized medicine is in the field of genomics. The human genome contains approximately 3 billion base pairs, representing an enormous amount of information. AI algorithms can sift through this data to identify genetic variants associated with diseases or responses to treatments. For instance, in oncology, AI can analyze a patient's tumor genome to predict how they will respond to different chemotherapy drugs, allowing oncologists to choose the most effective treatment with fewer side effects.

Another promising area is pharmacogenomics, the study of how genes affect a person's response to drugs. AI-driven pharmacogenomic tools can transform how medications are prescribed, moving away from the standard dosage recommendations. For example, some individuals metabolize certain drugs faster than others due to genetic differences, necessitating higher or lower doses to achieve the desired therapeutic effect. AI can analyze a patient's genetic profile to determine the optimal drug and dosage, minimizing trial and error and accelerating recovery.

Moreover, personalized medicine extends beyond treatment to the realm of preventive care. AI can predict the likelihood of developing diseases based on genetic, environmental, and lifestyle factors. This predictive capability allows for early interventions, potentially preventing diseases from manifesting at all. For example, individuals with a family history of diabetes can benefit from AI-driven tools that assess their risk and suggest personalized lifestyle changes to mitigate it.

Integrating lifestyle data is another significant advantage of AI in personalized medicine. Wearable devices and health apps generate a continuous stream of data related to physical activity, diet, sleep patterns, and other lifestyle factors. AI can analyze this data in real-time, offering insights that can guide personalized health advice. For example, if a wearable detects irregular heart rhythms, AI can alert the user to seek medical advice, potentially averting a serious health event.

AI also plays a crucial role in managing chronic conditions, such as diabetes, hypertension, and cardiovascular diseases. By continuously monitoring patients' health metrics and analyzing trends, AI can predict exacerbations and recommend timely interventions. This proactive approach can significantly improve the quality of life for individuals with chronic conditions, reducing hospitalizations and healthcare costs.

The use of AI in personalized medicine is not without its challenges. One major concern is data privacy. Personalized medicine relies on sensitive data, including genetic information and health records. Ensuring the security of this data is paramount to gaining patient trust and adhering to regulatory requirements. AI systems must be designed with robust encryption and compliance mechanisms to protect patient confidentiality.

Additionally, the integration of personalized medicine into mainstream healthcare requires significant changes in infrastructure and training. Healthcare providers must be equipped with the tools and knowledge to interpret AI-driven insights effectively. This transition represents a paradigm shift in how medical professionals approach diagnosis and treatment, emphasizing the importance of interdisciplinary collaboration between data scientists, geneticists, and clinicians.

Despite these challenges, the potential benefits of personalized medicine are too significant to ignore. The ongoing advancements in AI technology, alongside growing investments in healthcare innovation, are accelerating the adoption of personalized approaches. As AI becomes more sophisticated, its algorithms will continue to refine the precision of personalized medicine, leading to even more tailored and effective treatments.

Looking ahead, collaboration between technology companies, healthcare providers, and regulatory bodies will be crucial to realizing the full potential of personalized medicine. Investments in research and development, coupled with policies that support innovation while safeguarding patient privacy, will pave the way for more widespread adoption. As personalized medicine becomes more accessible, it has the potential to revolutionize healthcare delivery, making it more proactive, efficient, and patient-centered.

In summary, personalized medicine represents a paradigm shift in healthcare driven by AI. By leveraging vast datasets and sophisticated algorithms, personalized medicine tailors treatments and preventive measures to individual patients, enhancing efficacy and reducing risks. While challenges exist, the ongoing advancements and collaborations in this field promise to transform healthcare, leading to better health outcomes and improved quality of life for individuals worldwide.

AI in Diagnostics

Artificial Intelligence (AI) is revolutionizing the field of diagnostics, bringing a level of precision and speed previously unimaginable. By leveraging complex algorithms and vast datasets, AI systems can detect patterns and anomalies that often elude human eyes. One of the most significant advantages that AI brings to diagnostics is its ability to process and analyze large volumes of data quickly and accurately, thereby transforming how diseases are detected and diagnosed.

Consider radiology, a field where AI has made substantial inroads. AI algorithms, especially those based on deep learning, can analyze medical images like X-rays, MRIs, and CT scans with remarkable accuracy. These systems are trained on millions of images, allowing them to identify even subtle differences between healthy and abnormal tissues. For example, AI can assist radiologists in identifying early signs of tumors, fractures, or infections that might be too minute for manual detection. This capability can lead to earlier interventions and potentially save lives.

Beyond radiology, AI's diagnostic prowess extends to pathology. Histopathology, the microscopic examination of tissue samples to diagnose diseases, can benefit significantly from AI. AI-powered image analysis tools can quickly sift through slides, highlighting areas of concern for the pathologist to review further. This not only speeds up the diagnostic process but also enhances the accuracy of diagnoses by reducing human error.

AI is also breaking new ground in the field of genomics. By analyzing genetic data, AI can help identify mutations and genetic markers associated with various diseases. For instance, AI algorithms can scan through sequences of DNA to pinpoint anomalies that might lead to conditions such as cancer or rare genetic disorders. This can pave the way for personalized medicine, where treatments and preventive measures are tailored to an individual's genetic makeup.

Electronic Health Records (EHRs) are another area where AI is making a difference. Traditionally, interpreting EHRs to make diagnostic decisions has been labor-intensive and prone to human error. AI can analyze vast amounts of patient data rapidly, identifying trends and correlations that might not be immediately apparent. For instance, AI algorithms can flag potential complications or co-morbid conditions based on a patient's history, guiding doctors to make more informed decisions.

AI's role in diagnostics isn't limited to analyzing images or genetic data. Natural Language Processing (NLP), a subset of AI focusing on understanding and processing human language, is being used to sift through clinical notes and records. These AI systems can extract relevant information such as symptoms, diagnosis, and treatment plans from unstructured text, making it easier for healthcare providers to get a comprehensive view of a patient's health status.

Point-of-care diagnostics is another area where AI is making waves. Advances in AI-powered portable devices allow for rapid diagnostic tests to be conducted in remote or resource-limited settings. These devices can process and analyze samples on-site, providing real-time results. For example, AI-enabled diagnostic kits can detect infectious diseases like malaria or tuberculosis quickly and accurately, facilitating timely treatments.

Early Detection: AI's ability to analyze large datasets means it can detect diseases at much earlier stages, often before symptoms appear. This early detection is crucial for conditions such as cancer, where early intervention can significantly improve outcomes.

Accurate Diagnoses: With its ability to spot patterns and anomalies, AI can provide highly accurate diagnoses, reducing the occurrences of misdiagnosis or delayed diagnosis. This is particularly beneficial in complex cases where human diagnosticians might struggle.

Augmenting Human Expertise: Rather than replacing healthcare professionals, AI acts as an invaluable tool that augments their capabilities. Radiologists, pathologists, and other specialists can use AI as a second pair of eyes, improving the overall accuracy and reliability of their diagnoses.

Efficiency and Speed: AI can process vast amounts of data almost instantaneously, providing quicker results than traditional methods. This speed is pivotal in emergency settings where time-sensitive decisions can mean the difference between life and death.

The potential of AI in diagnostics goes beyond what we can currently see. Ongoing research and development are exploring how AI can integrate with other technologies such as the Internet of Things (IoT) and blockchain to provide even more robust diagnostic solutions. Imagine a world where wearable devices continuously monitor health parameters and alert both patients and doctors to any concerning changes. Or where blockchain ensures the secure and seamless sharing of diagnostic data between different healthcare providers globally.

Addressing limitations and challenges is equally crucial as we continue to integrate AI into diagnostics. One significant challenge is ensuring the ethical use of AI—a topic we'll explore further in the chapter on ethical considerations. There are concerns about data privacy, the potential for algorithmic bias, and the need for transparency in how AI systems make diagnostic decisions.

It's also essential to consider the infrastructure needed to support AI in diagnostics. High-quality data is the backbone of any AI-powered diagnostic tool. Hence, healthcare institutions must invest in digitizing records and ensuring data quality. Additionally, there's a need for ongoing training and education for healthcare professionals to familiarize them with AI tools and methodologies.

AI in diagnostics isn't just a technological advancement; it represents a paradigm shift in how we approach healthcare. By harnessing the power of AI, we can move towards a more proactive and precise healthcare system, where diseases are caught early, treatments are personalized, and healthcare professionals are empowered with the tools they need to make informed decisions. This shift promises not just to improve individual health outcomes but also to enhance the efficiency and effectiveness of healthcare systems worldwide.

Imagine a future where diagnostic errors are nearly obsolete, where the time between symptom onset and diagnosis is drastically shortened, and where every patient's treatment plan is optimized based on their unique biological makeup. AI is not just a tool but a catalyst propelling us towards this future, transforming the diagnostic landscape in ways that truly enhance our quality of life.

Health Monitoring Devices

Health monitoring devices are revolutionizing the way we manage our wellbeing, thanks to advancements in artificial intelligence (AI). These devices, which range from wearable fitness trackers to sophisticated home health monitors, are transforming healthcare from a reactive discipline into a proactive one. By continuously collecting and analyzing health data, they provide users with real-time insights that can prompt lifestyle changes, signal warnings of potential issues, and even offer critical information to healthcare providers. It's an exciting fusion of technology and healthcare that has the potential to improve quality of life and health outcomes significantly.

One of the most recognizable and widely used health monitoring devices is the wearable fitness tracker. Brands like Fitbit, Apple Watch, and Garmin have become household names by offering consumers the ability to track their physical activity, heart rate, sleep patterns, and more. These devices leverage AI algorithms to interpret the raw data they collect, providing actionable advice tailored to each user. For instance, they can identify irregularities in heart rate that might suggest a health issue, or suggest optimal times for physical activity based on previous performance and rest patterns.

In addition to fitness trackers, smartwatches have advanced their capabilities to include more comprehensive health monitoring features. For example, the latest models of the Apple Watch can perform electrocardiograms (ECGs) and detect potentially dangerous heart conditions such as atrial fibrillation. The integration of AI in these devices allows for continuous health monitoring with a level of accuracy and convenience that previously required frequent visits to a healthcare provider.

Beyond wearables, at-home health monitoring devices are becoming increasingly sophisticated. Devices like digital blood pressure monitors, glucose meters for diabetics, and smart thermometers are now equipped with AI to provide more precise and insightful readings. Some smart scales can measure body composition and track changes over time, offering users a more holistic view of their health.

One area where AI-powered health monitoring devices are making tremendous strides is in the management of chronic conditions. For instance, continuous glucose monitors (CGMs) for diabetics use AI to analyze blood sugar levels in real-time, sending alerts and recommendations to prevent dangerous spikes or drops. This can not only improve daily management but also provide data that healthcare providers can use to tailor treatment plans more effectively.

AI is also enhancing remote patient monitoring (RPM) technologies, which have gained significant attention, especially during the COVID-19 pandemic. These systems use a variety of sensors and devices to collect medical and health data from patients in real-time, then analyze the data to provide insights and alerts to both patients and healthcare providers. This continuous monitoring can help manage chronic diseases, prevent hospital readmissions, and reduce the burden on healthcare systems by keeping patients healthier at home.

Pregnancy and infant care have also seen benefits from AI-powered health monitoring devices. Wearable devices designed for expectant mothers can track fetal health metrics and maternal vitals, alerting them to any abnormalities. Postpartum, smart baby monitors equipped with AI can help parents keep track of their newborn's breathing, sleep patterns, and overall health.

The data collected by these devices is significant not only for individual users but also for broader public health initiatives. Aggregated, anonymized data can provide valuable insights into health trends and disease patterns, helping public health officials to respond more effectively to potential outbreaks and health issues. This is particularly crucial for managing and predicting the spread of infectious diseases, where early detection and intervention can save lives.

Furthermore, AI-driven health monitoring devices are promoting more engaged and informed patients. Users are empowered to understand their own health data and take a more active role in managing their wellbeing. This can lead to better adherence to treatment plans, healthier lifestyle choices, and a reduced need for emergency medical interventions.

Privacy and data security are of paramount importance when it comes to health monitoring devices. Ensuring that personal health data is secure and that users have control over who accesses their information is crucial. AI developers and device manufacturers must adhere to strict standards to maintain user trust and safeguard sensitive data.

The integration of AI in health monitoring devices is a testament to the transformative potential of technology in healthcare. As these devices continue to evolve, we can expect even more sophisticated and personalized health management solutions. The future might hold even more advanced biometric sensors, machine learning algorithms that can predict health issues before they become serious, and seamless integration with other health technologies to provide a comprehensive view of an individual's health.

In conclusion, health monitoring devices powered by artificial intelligence are not just gadgets or accessories; they are tools that can significantly enhance our understanding and management of personal health. By continuously collecting and analyzing health data, these devices provide valuable insights that can lead to early detection of potential issues, better management of chronic conditions, and overall improved health outcomes. As technology continues to advance, the potential for even greater innovations in health monitoring is immense, promising a future where AI plays a central role in maintaining and improving global health.

Chapter 4:
AI in Education

The integration of AI in education is revolutionizing the way we learn and teach, making the experience more personalized and efficient. Adaptive learning technologies are at the forefront, tailoring educational content to meet the unique needs of each student, ensuring no one is left behind. Virtual classrooms are breaking down geographical barriers, allowing students from around the world to access top-notch education and interactive learning experiences. AI tutors, acting as personal academic guides, provide real-time feedback and support, helping students grasp complex concepts with ease. This blend of technology and education not only enhances learning outcomes but also prepares students for a future where digital fluency is paramount. Through these innovations, AI is not just transforming education; it's democratizing it, empowering learners of all ages and backgrounds to achieve their fullest potential.

Adaptive Learning Technologies

Adaptive learning technologies are revolutionizing the educational landscape, bringing a more personalized and efficient way to learn. By utilizing data-driven algorithms and machine learning, these technologies can tailor educational experiences to meet the unique needs and learning styles of each student. This approach marks a significant departure from the traditional one-size-fits-all education model, paving the way for a more customized and enjoyable learning journey for everyone.

Imagine a classroom where each student receives personalized assignments, feedback, and instructional materials that adapt in real-time to their comprehension levels. This is the promise of adaptive learning technologies. Leveraging vast amounts of data, these systems analyze students' interactions with the material—such as which questions they get right or wrong, how long they take to answer, and even the specific mistakes they make. By doing so, they can dynamically adjust the difficulty and type of content provided, ensuring optimal learning paths for every individual. It's a far cry from the static textbooks and generalized worksheets that many of us grew up with.

One of the most compelling aspects of adaptive learning technologies is their potential to level the playing field for students with diverse learning needs. Whether a student struggles with dyslexia, attention deficit disorder, or another learning obstacle, adaptive systems can offer a tailored experience that accommodates their unique challenges. For instance, a student struggling with reading comprehension might be provided with additional resources and exercises to reinforce concepts, while a more advanced student could be challenged with higher-level material to keep them engaged.

This personalization extends to educators as well. Adaptive learning platforms often come equipped with dashboards that provide teachers with real-time insights into each student's progress and areas where they might need extra support. This allows teachers to intervene more effectively, offering targeted help rather than blanket remedial sessions. It turns the static role of educators into a dynamic, data-informed practice, where teachers can make more informed decisions to foster a conducive learning environment.

Yet, the benefits of adaptive learning technologies are not confined to K-12 education; they offer substantial advantages in higher education and professional development contexts. In universities, adaptive systems can help manage the broad range of student capabilities and backgrounds commonly found in lectures and seminars. On the professional development front, such technologies can assist companies in delivering customized training modules to employees, ensuring that each individual gets the specific skills and knowledge relevant to their roles.

The role of artificial intelligence in these adaptive learning technologies is immense. AI algorithms work tirelessly in the background, sifting through mountains of data to discern patterns and trends. Machine learning models continuously improve from these patterns, getting better at predicting what kind of content will be most beneficial for each user. Natural language processing (NLP) plays a pivotal role too, helping to understand and interpret text-based answers, and even enabling AI tutors to engage with students in more conversational ways. The potential for AI to not just educate but to understand and communicate with students more naturally is both fascinating and inspiring.

Motivationally, adaptive learning technologies can also enhance student engagement and motivation. When students see that their educational experience is tailored to their abilities and interests, it can make the process of learning feel more relevant and stimulating. Gamification elements are often integrated into these platforms, providing rewards, badges, and other incentives that make learning a more interactive and enjoyable experience. This kind of engagement can be particularly effective for younger students who are accustomed to interactive digital environments.

An inspiring example of adaptive learning in action is the success story of a public school in California. Implementing an adaptive mathematics program, the school saw a significant improvement in student performance. Scores in standardized tests began to rise, and teachers reported higher engagement levels. Students who were once intimidated by complex math problems found themselves gradually mastering content, buoyed by a system that responded to their individual struggles and successes. Such stories highlight the transformative potential adaptive learning holds for education systems worldwide.

However, it's important to acknowledge the challenges that come with adopting adaptive learning technologies. For one, the initial implementation costs can be high. Schools and educational institutions need to invest in the necessary hardware and software, and there's also the cost of training staff to effectively use these systems. Moreover, issues related to data privacy and security are paramount. With so much personal data being collected and analyzed, ensuring that this information is protected against breaches and misuse is critical.

Additionally, while adaptive learning technologies offer a personalized learning experience, they should complement—and not replace—human interaction in education. Teachers, mentors, and peers play irreplaceable roles in the learning process, providing emotional support, encouragement, and real-world context that machines currently cannot replicate. Balancing technology with human elements is crucial to creating a well-rounded, holistic educational experience.

In terms of future prospects, the field of adaptive learning technologies stands on the cusp of even greater advancements. As artificial intelligence continues to evolve, these systems will become increasingly sophisticated, capable of providing even more detailed and accurate personalization. Emerging technologies like augmented reality (AR) and virtual reality (VR) could also be integrated into adaptive learning platforms, offering immersive educational experiences that were once the realm of science fiction.

The transformative power of adaptive learning technologies can't be overstated. They're not just tools but enablers—allowing students to unlock their full potential, teachers to reach new heights in instructional effectiveness, and institutions to offer more inclusive and effective educational experiences. As we navigate the complexities and promises of AI in education, adaptive learning stands out as a beacon of what the future of education can look like: personalized, engaging, and profoundly impactful.

virtual classrooms

The landscape of education has been undergoing a significant transformation with the advent of Virtual Classrooms powered by Artificial Intelligence. Central to this revolution is the capability to break geographical barriers, allowing students and educators to connect from anywhere in the world. Virtual Classrooms leverage AI to create more immersive, personalized, and efficient learning environments.

In traditional educational settings, logistical constraints often hinder the accessibility and quality of education. With AI-driven Virtual Classrooms, these constraints are significantly reduced. Imagine a classroom where students in remote areas access the same quality of education as those in urban centers. This democratization of education is not just a futuristic dream but a present-day reality driven by AI technology.

AI in Virtual Classrooms provides a multitude of tools that enhance both teaching and learning experiences. One such tool is real-time translation, which breaks down language barriers instantly. For instance, a student in Japan can attend a lecture delivered in English and have it translated in real-time into Japanese, enabling seamless communication and understanding. AI algorithms analyze speech, detect nuances, and ensure that the translation maintains the same context and intent.

Virtual Classrooms powered by AI also offer intelligent tutoring systems that provide on-the-spot assistance. These AI tutors can monitor students' progress, identify areas where they struggle, and offer customized support. With AI's ability to analyze vast amounts of data quickly, it can assess a student's strengths and weaknesses much faster than a human tutor. This means that help is given exactly when and where it's needed, making the learning process more efficient and effective.

Moreover, AI-driven Virtual Classrooms employ personalized learning paths tailored to each student's unique needs. These paths are designed based on continuous assessments, ensuring that students progress at their own pace. The adaptive learning technologies underlying these systems analyze a student's performance in real-time and adjust the complexity of the material accordingly. This adaptive mechanism is pivotal in maintaining student engagement and promoting better learning outcomes.

Interactive elements are another cornerstone of AI-powered Virtual Classrooms. These elements include AI-driven simulations, virtual labs, and gamified learning experiences. For example, in a virtual biology class, students can interact with a digital model of a cell, manipulate its components, and observe the outcomes. Such interactive, hands-on experiences not only make learning more engaging but also reinforce theoretical knowledge through practical application.

AI also enhances collaboration in Virtual Classrooms. AI-facilitated discussion forums and study groups can form based on students' interests and academic needs. These intelligent groupings foster a collaborative learning environment where peer-to-peer interactions are optimized. AI systems can track the interactions and participation levels of students, providing feedback to both students and instructors. This data-driven insight helps in identifying students who may need additional support or those who are excelling and can help peers.

Another transformative aspect of AI in Virtual Classrooms is its role in administrative tasks. AI can automate scheduling, attendance tracking, and grading, freeing up educators to focus on teaching and mentoring. Automated grading systems analyze submissions, offering detailed feedback and identifying common errors among students. This constant formative assessment aids students in continuous improvement, turning each assessment into a learning opportunity rather than just an evaluation.

While these technological advancements are impressive, the benefits of AI in Virtual Classrooms extend beyond academic performance. These classrooms can support emotional and social learning by including AI tools that monitor students' emotional states. Using facial recognition and sentiment analysis, these systems can identify when a student appears disengaged or stressed and alert the instructor to provide appropriate support. It's an aspect of AI that adds a layer of empathy to digital learning, making the virtual environment feel more human.

AI can also facilitate accommodating diverse learning styles and needs. Students with disabilities benefit significantly from AI technologies that provide accessible learning materials. For example, AI-driven screen readers, speech-to-text converters, and customized user interfaces ensure that all students, regardless of their physical or cognitive abilities, can participate fully in the classroom.

Virtual Classrooms also present opportunities for professional development for educators. AI can analyze teaching patterns and provide insights on areas for improvement. It allows instructors to receive constructive feedback on their methodologies and adopt best practices. Furthermore, virtual professional learning communities can form where educators globally share resources, strategies, and support each other, fostering a culture of continuous professional growth.

Privacy and security are significant considerations in AI-powered Virtual Classrooms. It's essential to implement robust measures to protect sensitive student data. AI can aid in this as well, using advanced encryption and monitoring techniques to safeguard information. These systems can detect and respond to potential security breaches in real-time, offering a layer of protection that instills confidence in parents, students, and educators alike.

The future of learning through Virtual Classrooms isn't confined to the immediate benefits. Long-term, we can expect further integration of AI with emerging technologies like augmented reality (AR) and virtual reality (VR). This could lead to even more immersive learning environments, where students can explore historical events, complex scientific phenomena, or distant planets in a simulated yet highly interactive space. The combination of AI with AR and VR stands to redefine experiential learning, making subjects come to life in unprecedented ways.

As we look ahead, it's crucial to recognize and mitigate the challenges associated with AI in Virtual Classrooms. Inclusivity remains a concern, as not all students have equal access to the necessary technology and the internet. Bridging this digital divide is vital to ensure equitable learning opportunities. Additionally, we must consider the ethical implications of AI, such as data privacy, bias in algorithms, and the role of human oversight in AI-driven educational tools. Navigating these challenges will require concerted efforts from educators, policymakers, technologists, and communities.

In embracing Virtual Classrooms, it's vital to maintain a balance between technology and the human element of education. AI should be viewed as an enhancer rather than a replacement. The role of educators remains irreplaceable; their empathy, mentorship, and ability to inspire are aspects that technology cannot replicate. Using AI as a support tool allows teachers to focus more on these irreplaceable elements, ultimately enriching the educational experience for students.

A final point to consider is the joy of learning that Virtual Classrooms can foster. When technology eradicates obstacles to education and caters to individual needs without compromising on quality, learning becomes an enjoyable pursuit rather than a chore. The AI-driven personalization, engagement, and support collectively foster a love for learning that paves the way for lifelong educational journeys.

In conclusion, AI in Virtual Classrooms represents a monumental shift in how we perceive and deliver education. Its potential to transform learning environments, personalize education, and break down barriers is just the beginning. As we witness this evolution, the challenge and the opportunity lie in harnessing AI's capabilities while prioritizing the human touch in education. The journey promises to be as transformative as it is exciting.

ai tutors

The promise of AI tutors revolutionizes education in ways we've only dreamt of. Traditional classroom environments offer limited instructor time, and it's nearly impossible for one teacher to tailor their approach to meet each student's unique needs. AI tutors change the game by providing personalized educational experiences, adapting to individual learning styles, and offering round-the-clock availability. This technology doesn't just supplement human educators; it creates entirely new ways of learning.

One of the hallmark features of AI tutors is their ability to adapt to the pace and learning style of each student. For instance, if a learner struggles to grasp a math concept, the AI can provide additional examples and problems to practice with. Conversely, for learners who excel, the AI can offer more challenging material to keep them engaged. It's a dynamic cycle of teaching tailored to individual needs that a single human instructor could never achieve on such a scale.

AI tutors can also give students instant feedback. In many educational settings, feedback on assignments or comprehension is delayed, sometimes impacting the learning process. With AI tutors, the feedback loop is immediate. This immediacy helps learners understand and rectify mistakes in real-time, fostering a more effective learning experience. Imagine a world where students never have to wait for their papers to be graded; they get the feedback they need right away.

In terms of availability, AI tutors stand unparalleled. Traditional tutoring sessions require coordination between the student's and teacher's schedules. AI tutors, however, are available anytime, anywhere. This 24/7 availability ensures that students can learn at their own convenience, making education more accessible than ever before. Students in different time zones or those with irregular schedules no longer need to miss out.

Moreover, AI tutors are increasingly bilingual or multilingual, breaking down educational barriers for non-native speakers. For example, English-speaking students learning French can practice their conversational skills with an AI tutor programmed to respond in French. This kind of language practice can significantly boost confidence and competence in a new language, something that's hard to achieve in standard classroom settings.

The implementation of AI tutors in education is not just limited to assisting students but also extends to supporting teachers. AI tutors can handle repetitive, time-consuming tasks such as grading and attendance, freeing up educators to focus on more complex aspects of teaching. They act as augmentative tools for teachers, allowing them to spend more time on personalized student interaction and lesson planning.

Interactive and engaging, AI tutors utilize gamification techniques to make learning fun. Students earn points, badges, and rewards as they progress through materials, a method proven to enhance motivation and engagement. These gamified elements make learning feel less like a chore and more like an adventure, particularly beneficial for younger students who might otherwise struggle to maintain interest.

One of the most exciting developments in AI tutoring is the capacity for data analytics. AI systems can track student progress over time, identifying strengths and weaknesses that may not be apparent through traditional methods. Teachers and parents can access detailed reports, offering insights that can help customize educational plans for each student. This data-driven approach makes education more effective by targeting areas that need improvement.

There's also a growing emphasis on emotional intelligence in AI tutors. Modern AI systems are being designed to recognize and respond to a student's emotional state. If a student shows signs of frustration, the AI can offer encouragement or suggest taking a break. This attention to emotional well-being is crucial, as it creates a more supportive and empathetic learning environment. Technology is moving beyond just being smart; it's becoming emotionally aware.

Virtual reality (VR) and augmented reality (AR) are another exciting frontier for AI tutoring. Imagine studying ancient civilizations and being able to take a virtual tour of the Roman Colosseum or the Pyramids of Giza. These immersive experiences make learning more dynamic and engaging, helping students retain information more effectively. The integration of VR and AR with AI tutors turns traditional learning methods on their head, making education a multidimensional experience.

Despite the numerous benefits, it's important to acknowledge potential challenges. There are concerns about data privacy, the digital divide, and the over-reliance on technology in the learning process. Policymakers, educators, and technologists must collaborate to address these issues, ensuring that the benefits of AI tutors are accessible to all students while safeguarding privacy and equity. Thoughtful implementation is the key to overcoming these hurdles.

Furthermore, incorporating AI tutors in public and private educational institutions requires investment in infrastructure and training.

Schools need access to high-speed internet, modern devices, and ongoing technical support to effectively utilize AI technologies. Teachers also need to be trained on how to integrate AI tutors into their curriculums. These logistics shouldn't be a barrier but rather a point of consideration for stakeholders aiming to revolutionize education.

In conclusion, AI tutors represent a monumental shift in how education is delivered. From personalized learning experiences and instant feedback to 24/7 availability and emotional intelligence, AI tutors have the potential to transform educational outcomes for students worldwide. As we continue to innovate and refine these technologies, the vision of a more accessible, effective, and engaging educational landscape comes ever closer to reality. Embracing these advancements thoughtfully and responsibly will ensure that the future of education is bright and inclusive.

CHAPTER 5:
AI IN PERSONAL FINANCE

A I in personal finance is revolutionizing the way we handle money, bringing unparalleled precision and efficiency to our financial lives. Imagine automated budgeting tools that analyze spending patterns and offer personalized savings tips, helping you to manage your finances without the hassle of manual tracking. AI-driven investment platforms are not just for the wealthy; they democratize investing by providing data-driven insights and portfolio management tailored to individual risk profiles. Meanwhile, advanced fraud detection systems continuously monitor transactions, enhancing security by quickly identifying and responding to suspicious activities. These innovations are not futuristic dreams; they're transforming financial wellness right now, encouraging smarter, safer financial decisions and empowering users to achieve their monetary goals with confidence and ease.

Automated Budgeting Tools

In today's world, managing personal finances has become more streamlined and efficient, thanks to the advent of Artificial Intelligence (AI). Automated budgeting tools, driven by AI, provide a comprehensive way to track income, monitor expenses, and ensure that financial goals are met with minimal effort. These tools are transforming how individuals interact with their banking and financial data, making it simpler to manage money while fostering better financial habits.

One of the primary benefits of automated budgeting tools is their ability to integrate with various financial accounts seamlessly. By connecting to bank accounts, credit cards, and investment portfolios, these tools can offer a holistic view of an individual's financial health. They categorize transactions automatically, eliminating the need for manual entry, and provide insights into spending patterns and areas where users can cut back or save more. This automation results in not only time savings but also in more accurate and timely financial analyses.

Moreover, many automated budgeting tools utilize AI to predict future expenses based on past spending habits. This predictive analysis helps users prepare for upcoming financial commitments and avoid potential shortfalls. For instance, if the tool identifies a recurring pattern of high electrical bills during the summer months, it can suggest setting aside extra funds in advance. Such foresight enables users to create more realistic and effective budgets, reducing financial stress and enhancing overall financial stability.

AI-powered budgeting tools also offer personalized financial advice. These recommendations are tailored based on an individual's unique financial situation, spending behaviors, and goals. Unlike generic advice found online or from traditional financial advisors, AI-driven suggestions are dynamic and evolve as the user's financial landscape changes. This bespoke advice ranges from suggesting savings opportunities and optimal debt repayment strategies to alerting users about unusual transactions that might indicate fraud.

Another advantage of these tools is their ability to set and track financial goals. Whether it's saving for a down payment on a house, planning a vacation, or building an emergency fund, AI-driven budgeting platforms allow users to set specific goals and monitor their progress. Visual progress charts and milestone notifications keep users motivated and on track, making the journey towards achieving financial objectives more engaging and less daunting.

Additionally, many automated budgeting tools feature AI-enhanced analytics to provide users with deeper insights into their financial behaviors. Users can visualize their spending in various categories such as dining, entertainment, and groceries. Interactive charts and graphs make it easy to identify which areas are consuming the most resources and where adjustments can be made. This level of granularity helps users understand their financial habits better and make informed decisions about their future spending.

Security is a paramount concern when it comes to personal finance, and automated budgeting tools leverage AI to enhance this aspect significantly. Through AI, these tools can detect unusual spending activities that may signify fraudulent transactions. Advanced algorithms analyze user behavior and flag any discrepancies, ensuring that users are promptly informed of any suspicious activity. As a result, individuals can respond quickly to potential threats, protecting their financial assets more effectively.

The convenience of having all financial information aggregated in one place cannot be overstated. AI-driven tools provide a centralized dashboard where users can view their net worth, track incomes and expenses, review financial goals, and receive actionable insights—all in real-time. This unified approach simplifies financial management and empowers users to take control of their finances without feeling overwhelmed.

It's also worth noting that these tools continuously learn and adapt. As AI technologies advance, so do the capabilities of automated budgeting tools. They become more adept at understanding individual preferences and financial behaviors, which results in increasingly accurate recommendations and forecasts. The learning algorithms can adjust the advice provided, making it ever more relevant to the user's evolving financial situation.

In the realm of personal finance, discipline and consistency are key to achieving long-term goals. Automated budgeting tools help instill these virtues by providing regular reminders and updates. Whether it's a notification about upcoming bills, an alert about exceeded budget limits, or a congratulatory message on reaching a savings milestone, these prompts keep users engaged and committed to their financial plans.

Furthermore, these tools often incorporate gamification elements to make financial management more enjoyable. Features such as earning badges for achieving savings targets or participating in budget challenges turn mundane financial tasks into engaging activities. This element of fun can motivate users to stick with their budgeting plans and foster a positive association with managing their finances.

The impact of automated budgeting tools extends beyond individual users. Families and households can also benefit significantly. Many of these platforms offer the capability to manage shared expenses and budgets, ensuring transparent and coordinated financial planning among family members. This feature is especially useful for couples or roommates, helping to avoid misunderstandings and ensuring that everyone is on the same page concerning financial matters.

Small business owners and freelancers can also leverage AI-driven budgeting tools to manage their finances more efficiently. These individuals often juggle personal and business expenses, making financial tracking more complex. Automated tools can separate and categorize expenses accordingly, providing clear insights into the financial health of both personal and business accounts. This dual functionality simplifies tax preparation, expense management, and financial forecasting, ultimately aiding in better business decisions.

As these tools become more advanced, integration with other AI-driven financial services is becoming commonplace. For instance, certain budgeting apps can connect with AI-driven investment platforms, allowing users to manage their budgets and investments from a single interface. This seamless integration ensures a cohesive financial strategy, enabling users to balance present financial needs with future investment goals.

The future of automated budgeting tools looks promising, with many exciting advancements on the horizon. Innovations such as voice-activated financial assistance through smart home devices, more robust predictive analytics, and enhanced real-time collaboration features are just a few of the developments poised to revolutionize personal finance further. Users can expect even more intuitive and user-friendly experiences as these technologies continue to evolve.

In conclusion, AI-powered automated budgeting tools are redefining personal finance by offering a smarter, more efficient way to manage money. Through seamless integration, predictive analysis, personalized advice, and enhanced security, these tools empower individuals to take control of their financial futures. The continuous advancements in AI ensure that these tools will only become more sophisticated, making financial management more accessible and enjoyable for everyone.

ai-driven investment platforms

The world of investment is undergoing a seismic shift with the introduction of AI-driven investment platforms. While traditional advisors and brokers still play significant roles, AI has brought a fresh wave of innovation that democratizes access to sophisticated investment strategies. These platforms act as personal financial advisors, employing complex algorithms and machine learning models to manage portfolios, forecast market trends, and even automate trades. For the everyday investor, this means more opportunities to grow their wealth with less effort and more accuracy.

AI-driven investment platforms are essentially a blend of technology and finance designed to give investors real-time insights into their portfolios. They analyze massive amounts of data to identify patterns and predict future movements in the stock market. This capability allows these platforms to offer personalized investment advice, fine-tuned to an individual's specific financial goals and risk tolerance. Whether you're a novice investor or a seasoned trader, these platforms make investment planning more accessible and efficient.

Among the most compelling features of AI-driven investment platforms is their ability to adapt in real time. As market conditions change, these systems can adjust your portfolio accordingly, ensuring that your investments are aligned with current economic trends. This is achieved through neural networks and predictive analytics, which offer a depth of market analysis that simply isn't feasible manually. Essentially, it's like having a highly skilled financial analyst working for you round the clock.

Another striking advantage is the cost-effectiveness of these platforms. Traditional investment advisory services can be prohibitively expensive, often putting them out of reach for ordinary individuals. AI-driven platforms, on the other hand, operate at a fraction of that cost, making high-quality investment advice accessible to everyone. The reduced costs are a result of automation and streamlined operations, which eliminate the need for large teams of analysts and clerks.

For those wary of entrusting their hard-earned money to a machine, transparency is a key aspect these platforms are increasingly adopting. Alongside recommendations, they often provide detailed explanations of how decisions are made, based on the data analyzed. This transparency helps build trust, ensuring that users understand the rationale behind each investment choice.

Consider the example of "robo-advisors," a term commonly used to describe these AI-driven platforms. They handle everything from rebalancing portfolios to tax-loss harvesting, activities that are crucial for maximizing returns but typically require expert intervention. Robo-advisors use algorithms to optimize these processes, reducing human error and emotional bias, which can often skew investment decisions. This results in a more disciplined and systematic approach to investing.

The inclusiveness provided by AI-driven platforms is also noteworthy. Previously, the most advanced and profitable investment strategies were often reserved for institutional investors and the wealthy. However, AI has leveled the playing field by making these strategies available to the average person. Techniques like algorithmic trading and risk parity portfolios can now be part of anyone's financial toolkit.

One of the standout benefits of AI in this domain is its ability to provide tailored advice. Traditional financial advisors might rely on generalized strategies that aren't always a perfect fit for every client. In contrast, AI can analyze individual financial profiles, account for personal milestones such as retirement or buying a home, and recommend investment plans that align with these life goals. This level of customization has elevated the quality of investment advice available to the broader public.

Furthermore, AI-driven investment platforms typically offer a user-friendly interface, making it easier for individuals to understand and manage their investments. These platforms often come with dashboards that provide snapshots of one's financial health, performance metrics, and future projections. This visualization makes complex financial data comprehensible, helping users make informed decisions without needing an in-depth knowledge of financial markets.

Machine learning algorithms are the backbone of these platforms. They continuously learn from historical data to improve their predictive capabilities. If a particular stock has a high probability of performing well based on past trends, the algorithm will recognize this pattern and recommend an appropriate action. This self-improving trait of AI ensures that the quality of advice keeps getting better over time.

Security is another critical aspect where AI-driven platforms excel. Advanced algorithms can detect unusual patterns of behavior, helping to protect investments from fraud and cyber threats. These platforms utilize sophisticated encryption and real-time monitoring to ensure that sensitive financial information remains secure. Fraud detection systems are continually learning from new data, making it harder for malicious activities to go undetected.

Let's also not overlook the role of sentiment analysis in AI-driven investment platforms. These systems can analyze news articles, social media posts, and even statements from financial analysts to gauge the market sentiment about a particular stock or the economy as a whole. Tools that were once the domain of hedge funds and institutional investors are now accessible to individual investors, arming them with insights that were previously unimaginable.

The advancements in natural language processing (NLP) contribute significantly to the effectiveness of AI in personal investment. NLP helps in interpreting unstructured data like news and financial reports, converting them into actionable insights. For instance, a sudden surge in online searches for a company's product might indicate a future rise in that company's stock. These nuances are automatically captured and analyzed to inform investment decisions.

In terms of long-term wealth generation, AI-driven platforms are constantly innovating to offer features that were inconceivable a decade ago. For instance, some platforms now include predictive analytics to forecast the performance of entire portfolios under different economic scenarios. Such insights enable users to prepare for market volatilities and make informed adjustments to their investment strategies accordingly.

AI-driven platforms also facilitate better diversification, which is crucial for risk management. Algorithms can identify correlations and causations among different assets, advising on the best mix to optimize returns while minimizing risks. This capability is particularly invaluable for those who may not have a deep understanding of financial theory but still wish to build a balanced portfolio.

Emotion is often the Achilles' heel of investing. Fear and greed can lead to irrational decisions that may harm an investor's long-term financial health. AI, devoid of emotional biases, makes decisions based solely on data and probability, ensuring a more rational and disciplined approach to investing.

In this rapidly evolving landscape, regulatory oversight remains a critical factor. As AI-driven platforms gain popularity, they are subject to stringent regulations designed to protect consumers. Compliance with these regulations not only ensures the security and reliability of these platforms but also builds trust among users wary of new technologies.

While AI-driven investment platforms are remarkable, they are not without their limitations. They are dependent on the quality of data they receive and the algorithms designed to process this data. Anomalies or data inaccuracies can lead to flawed recommendations. Hence, continuous monitoring and updating of these systems are essential to maintain their effectiveness.

The integration of AI in personal finance, particularly in investment platforms, signals a transformative era where state-of-the-art technology allows anyone to partake in sophisticated investment strategies that were once the exclusive domain of the financial elite. The result is an investment landscape that is more inclusive, transparent, and accessible, offering unprecedented opportunities for wealth creation and financial security.

In conclusion, AI-driven investment platforms represent a paradigm shift in how we approach investments. Their ability to process vast amounts of data, predict market trends

Fraud Detection

In the realm of personal finance, fraud detection is an area where Artificial Intelligence (AI) has made enormous strides. Financial fraud is a persistent issue worldwide, affecting individuals and institutions, and costing billions of dollars annually. AI presents a transformative solution to this problem by offering advanced methods for monitoring, analyzing, and predicting fraudulent activities in real time.

Traditional methods of fraud detection relied on static rules and historical patterns, which were often slow to adapt to new tactics employed by fraudsters. AI, on the other hand, leverages machine learning algorithms that can adapt and evolve as new types of fraud emerge. This flexibility allows AI systems to recognize anomalous transactions that deviate from a user's typical behavior, flagging them for further investigation.

The core of AI-driven fraud detection is machine learning. Machine learning models can be trained on vast datasets that include both fraudulent and legitimate financial transactions. These models learn to distinguish between the two, identifying subtle patterns that may be invisible to human analysts. Over time, as more data is fed into the system, the models become increasingly accurate.

Another critical aspect of AI in fraud detection is anomaly detection. AI systems continuously monitor transactions for irregularities that could indicate fraudulent activity. For instance, if a credit card that is typically used within the United States is suddenly used to make a purchase in a country miles away, the AI system can flag this as potentially suspicious. The system can then prompt additional verification steps before approving the transaction.

AI also enhances fraud detection through predictive analytics. By analyzing historical transaction data, AI can predict the likelihood of future fraudulent activities. Predictive models assess risk based on variables such as transaction amount, location, time, and frequency. By forecasting potential fraud, these models help financial institutions and individuals take proactive measures to mitigate risks.

Natural Language Processing (NLP) plays a significant role in AI-powered fraud detection. NLP algorithms can analyze unstructured data such as emails, social media posts, and customer service interactions to detect signs of fraud. For example, an AI system can identify language patterns indicative of phishing attempts or social engineering tactics, providing an additional layer of protection.

Beyond detection, AI is also crucial in mitigating the impact of fraud. Once fraudulent activity is identified, AI systems can automate the immediate response, such as temporarily freezing accounts, alerting customers, and notifying law enforcement. These rapid responses are essential in preventing extensive damage and recovering lost funds.

AI's role in fraud detection is not limited to financial transactions alone. Insurance fraud is another significant area where AI proves invaluable. By examining claims data and cross-referencing it with other sources, AI can identify inconsistencies and patterns that suggest fraudulent claims. This capability allows insurance companies to process legitimate claims more efficiently while reducing payouts on fraudulent ones.

While AI offers robust solutions for fraud detection, it is not without challenges. One of the primary concerns is the potential for false positives, where legitimate transactions are mistakenly flagged as fraudulent. This issue can lead to customer frustration and inconvenience. However, as AI systems continue to learn and improve, their accuracy in distinguishing between fraudulent and legitimate activities is expected to increase.

Another challenge is the ethical use of AI in fraud detection. It is essential to ensure that AI systems do not inadvertently discriminate against certain groups of people. Bias in training data can result in unfair targeting of specific demographics. Therefore, it is crucial for developers to implement measures that ensure the fairness and transparency of AI algorithms.

Despite these challenges, the benefits of AI in fraud detection far outweigh the potential drawbacks. Financial institutions and personal finance platforms that adopt AI-driven fraud detection systems are better equipped to protect their customers and assets. The real-time monitoring, adaptive learning, and predictive capabilities of AI provide a formidable defense against ever-evolving fraudulent tactics.

Moreover, AI's ability to handle large volumes of data at high speed makes it a powerful tool for fraud prevention. Human analysts, no matter how skilled, cannot match the efficiency and accuracy of AI systems in processing and analyzing data. As such, AI serves as an invaluable ally in the fight against financial fraud.

In conclusion, AI's contribution to fraud detection in personal finance is revolutionary. By leveraging machine learning, anomaly detection, predictive analytics, and NLP, AI systems provide a comprehensive and dynamic approach to identifying and mitigating fraud. While there are challenges to address, the continued advancement and refinement of AI technologies hold great promise for enhancing financial security and fostering trust in digital financial systems. As we look to the future, the integration of AI in fraud detection will undoubtedly become even more sophisticated, providing even greater protection for individuals and institutions alike.

CHAPTER 6:
AI IN ENTERTAINMENT

Artificial Intelligence is revolutionizing the entertainment industry, infusing it with unprecedented levels of personalization and interactivity. Streaming services like Netflix and Spotify use AI algorithms to analyze viewing and listening habits, providing tailored recommendations that align perfectly with user preferences. In the realm of video games, AI opponents have progressed from predictable and repetitive to highly adaptive and challenging, creating immersive experiences that keep players engaged. Virtual reality, too, is enhanced by AI, making virtual worlds more responsive and lifelike. These innovations not only elevate user satisfaction but also open new frontiers in storytelling and content creation. Ultimately, AI's influence in entertainment is democratizing access to diverse content, enriching our leisure time, and transforming how we consume media.

Streaming Recommendations

As we delve into the world of "AI in Entertainment," the realm of streaming recommendations stands as a prime example of AI's transformative impact. Streaming services like Netflix, Hulu, and Amazon Prime rely heavily on AI algorithms to tailor content recommendations to individual users. These algorithms sift through colossal amounts of data to predict what you might enjoy watching next, creating a deeply personalized viewing experience.

The recommendation systems are like sophisticated digital match-makers, meticulously analyzing your past behaviors, viewing history, and preferences to present you with a curated list of TV shows, movies, and documentaries. They don't just consider what you watched, but also how you watched it—whether you binged an entire series in one night or spread it over several weeks. This level of detail enables AI to suggest titles with uncanny accuracy, making it seem as though the platform knows your tastes and preferences intimately.

An average user might not realize just how intricate these algorithms are. They incorporate machine learning techniques such as collaborative filtering, content-based filtering, and hybrid methods to offer accurate recommendations. Collaborative filtering relies on the collective behavior of users, suggesting content based on the similarities between your viewing habits and those of others. If someone with a similar taste enjoyed a particular show, the algorithm assumes you might as well.

On the other hand, content-based filtering takes a different approach. It focuses on the attributes of the content itself, like genre, actors, and even minute details such as soundtrack and cinematography style. By examining what you've watched and liked, this method can recommend other content that shares the same characteristics. Combining both collaborative and content-based filtering, hybrid methods leverage the strengths and mitigate the weaknesses of each, offering the most balanced recommendations.

These algorithms are continually improving, thanks to advances in AI and machine learning. One such advancement is the use of deep learning techniques, where artificial neural networks mimic the human brain's ability to learn from data. These highly sophisticated networks can recognize patterns and make connections that simpler algorithms might miss. For example, they might catch subtle correlations between your favorite genres and the days of the week when you prefer to watch certain types of content, refining suggestions even further.

Some of the most captivating AI-driven recommendation engines also incorporate Natural Language Processing (NLP) to understand and analyze reviews, comments, and ratings. NLP enables the system to discern the sentiment behind user reviews, grouping similar movies or shows based on audience reactions. This can make recommendations richer and more nuanced, offering shows that generated similar viewer experiences.

Additionally, AI-driven recommendations aren't confined to video streaming alone. Music platforms like Spotify and Apple Music employ similar algorithms to tailor playlists and song suggestions. They use a meticulous blend of your listening history, song attributes, and even temporal factors to craft playlists that resonate with you, sometimes introducing you to new genres or artists that you might never have discovered otherwise.

Smart recommendations extend beyond mere convenience; they actually enhance the entire entertainment experience. They create an ecosystem where discovering new content feels less like a chore and more like a serendipitous adventure. When done correctly, AI recommendations can introduce viewers to hidden gems they would have never found on their own, leading to more diverse and enriched entertainment consumption.

This personalized approach also benefits streaming services by increasing user engagement. When users find relevant content quickly, they spend more time on the platform. Increased engagement translates to higher retention rates and, ultimately, greater revenue for these companies. The more data they collect about user preferences, the more accurately the algorithms can function, creating a virtuous cycle of continuous improvement.

Despite its advantages, streaming recommendations powered by AI aren't without challenges. One significant issue is the potential for "filter bubbles," where the algorithms keep suggesting similar types of content, limiting exposure to a more diverse array of genres and subjects. Users might find themselves trapped in a loop, watching only what the algorithm believes they want to see. To mitigate this, some platforms are introducing features that encourage serendipitous discovery by occasionally offering random or popular choices outside your usual preferences.

Privacy concerns also loom large. These algorithms require extensive data to function effectively, raising questions about how user information is collected, stored, and utilized. Transparency and user control over data are pivotal in addressing these concerns. Some services now offer more robust settings where users can manage their data preferences, turning off certain types of tracking if they wish.

Going forward, the role of AI in streaming recommendations is only set to grow. Emerging technologies such as augmented and virtual reality promise new dimensions of personalized content. Imagine an AI recommending not just a movie but an entire immersive experience tailored to your tastes, complete with personalized interactive elements. This is not just the future of entertainment but the future of truly personalized experiences.

As AI becomes more integrated into our lives, its role in shaping our entertainment choices will continue to evolve. There's a delicate balance to maintain—between harnessing the power of data to deliver personalized experiences and ensuring that those experiences remain broad and varied enough to keep life's serendipities alive. If struck well, this balance could redefine how we experience entertainment in the digital age.

Ultimately, AI-driven streaming recommendations offer a fascinating glimpse into the future, where technology meets personalization to create richer, more engaging experiences. As these systems become more sophisticated, they will not only entertain but also inspire, educate, and connect us in ways we can only begin to imagine. So, as you queue up your next show or playlist, remember—there's an incredibly smart system working behind the scenes, dedicated to making your moments of leisure truly enjoyable.

AI in Video Games

Artificial Intelligence has had an extraordinary impact on the video game industry, transforming how games are developed and experienced. Whether it's creating more believable non-playable characters (NPCs), generating landscapes, or personalizing gameplay, AI techniques are making games more immersive than ever. Imagine playing an RPG where NPCs adapt to your actions, remember past interactions, and dynamically change their behavior. AI enables that level of complexity, turning static game worlds into interactive environments with depth and richness.

One of the most exciting aspects of AI in video games is procedural content generation. Instead of designers crafting every aspect of a game manually, AI algorithms can generate vast landscapes, intricate dungeons, and even entire game worlds on the fly. This allows for virtually infinite variability and replayability. For example, the "roguelike" genre relies heavily on procedural generation, ensuring no two playthroughs are ever the same. Not only does this keep players engaged, but it also significantly reduces the time and cost associated with game development.

Machine learning algorithms can learn from player behavior to adjust game difficulty in real time, making games more accessible to novices while still challenging seasoned players. This dynamic difficulty adjustment ensures that players experience a balanced level of challenge, keeping frustration at bay and maintaining the fun. Games like

Left 4 Dead employ AI directors that monitor the players' performances and adjust the game's intensity by altering the number and difficulty of enemies, providing a tailored experience each time.

AI in video games also brings a new level of emotional engagement through advanced facial recognition and emotional AI. Characters in games can now exhibit complex emotional states and respond to the player's actions and decisions. Quantic Dream's Detroit: Become Human, for instance, features characters whose emotional responses are generated by AI, making interactions feel genuine and impactful. This emotional depth adds layers of meaning to a player's journey, making each decision and its consequences feel more significant.

The scope of AI extends beyond the gameplay experience itself; it's also changing how games are tested and debugged. Automated testing powered by AI allows developers to identify and fix bugs far more efficiently than traditional methods. AI-driven bots run through numerous permutations of gameplay, highlighting potential problems that human testers might miss. This results in a more polished final product, released to players in a shorter time frame.

Moreover, AI is revolutionizing multiplayer gaming. Intelligent matchmaking algorithms analyze player behavior, skill levels, and preferences to ensure fair and enjoyable matches. By doing so, AI helps create balanced teams and thrilling contests, enhancing the overall multiplayer experience. Services like Microsoft's TrueSkill utilize machine learning to continuously refine player rankings, contributing to more competitive and engaging gameplay for everyone involved.

The development of AI companions and assistants in video games is another noteworthy innovation. These virtual aides can guide players through complex quests, provide hints without spoiling the game, and even assist in combat scenarios. In games like The Legend of Zelda: Breath of the Wild, AI companions analyze the player's situation and offer advice or assistance at critical moments, making the experience smoother and more enjoyable.

AI's role in video games isn't limited to player-facing features; it also assists developers in the creative process. AI-driven tools can generate artwork, model assets, and even compose background music, allowing creative teams to focus on other aspects of game design. These tools can learn from existing game assets to create new content that matches the desired style and feel, significantly expediting the creative process. Ubisoft's game development tool, Ghostwriter, assists in generating dialogue, reducing the workload of writers and enabling them to explore more complex narratives.

Voice recognition and natural language processing technologies are making their way into video games, allowing players to interact with game worlds and characters using natural speech. Imagine giving voice commands to your squad in a tactical shooter or negotiating with an NPC by speaking into your microphone. AI-driven voice interaction offers a more intuitive and immersive way to play, bringing games closer to the realms of interactive storytelling.

AI in video games is continuously evolving, with newer techniques such as deep learning and reinforcement learning pushing the boundaries of what's possible. These advanced algorithms allow game characters to learn and improve over time, adapting to players in ways never before possible. Reinforcement learning, in particular, enables AI to develop strategies and tactics by playing the game repeatedly and learning from each round. The result is more unpredictable, and therefore realistic and challenging AI opponents.

Looking ahead, the future of AI in video games promises even more groundbreaking innovations. As AI continues to advance, we can anticipate games with even greater levels of interactivity, dynamism, and realism. Augmented reality (AR) and virtual reality (VR) combined with AI can offer fully immersive worlds where the virtual and real blend seamlessly, enhancing the gaming experience to unprecedented levels. Imagine AI-driven characters in a VR game that not only respond to your actions but can also move about and interact with the real-world environment around you.

Additionally, AI could enable multiplayer games to host larger, more complex worlds with thousands of players interacting simultaneously. Cloud gaming services, powered by AI, can manage immense computational loads and provide a seamless experience regardless of a player's hardware. By offloading processing to powerful servers, these services can deliver high-quality graphics and lifelike simulations even on less capable devices.

The ethical implications of AI in video games also warrant consideration. AI algorithms could be designed to monitor and influence player behavior, raising questions about privacy and the potential for manipulation. Developers must navigate these ethical waters carefully, ensuring that AI enhances the player experience without compromising on ethical standards. Transparent AI systems and user consent will become increasingly important as these technologies advance.

To sum up, the integration of AI in video games marks a substantial leap forward in how these virtual worlds are crafted and experienced. From procedural content generation and dynamic difficulty adjustments to emotional AI and voice interaction, the possibilities are virtually limitless. Video games are no longer just about passively navigating through pre-set scenarios; they are becoming adaptive, interactive experiences that can react and grow alongside the player. This transformation is setting the stage for the next generation of gaming, offering endless opportunities for innovation and engagement.

Virtual Reality Experiences

Imagine stepping into a world where reality bends and molds to your every whim. This is the promise of Virtual Reality (VR), an emerging frontier in the entertainment industry significantly enhanced by Artificial Intelligence (AI). As VR technology has evolved, its integration with AI has opened up a plethora of new opportunities for creating immersive experiences like never before.

At its core, a Virtual Reality experience aims to transport the user into an entirely different environment. When paired with AI, the realism and interactivity levels are drastically heightened. AI algorithms can adapt virtual environments in real time based on user interactions, making each experience unique and personalized. For example, in a VR gaming scenario, NPCs (non-playable characters) may react genuinely to player actions, thanks to sophisticated AI models analyzing behavioral patterns.

But its applications go far beyond gaming. Imagine a VR experience where you could walk through ancient ruins, interact with historical figures, or even learn surgery techniques. AI's role in these initiatives is pivotal. Advanced data analysis, pattern recognition, and predictive algorithms enable real-time corrections and guidance, making educational VR applications profoundly impactful.

One of the most exciting prospects of AI-enhanced VR experiences is in the realm of social interaction. Virtual social spaces, powered by AI, can offer genuine human-like interactions. AI-driven avatars can mimic human behavior and emotions, providing a more authentic experience. This is particularly beneficial in scenarios like virtual meetings or social gatherings, where conveying subtle emotional cues can make interactions more engaging and meaningful.

Even the realms of art and culture are not left untouched by this synergy. Museums and art galleries are leveraging AI in VR to offer virtual tours, curated by AI-driven guides who provide detailed, personalized narratives based on the visitor's interests and previous engagements. Imagine standing in front of Van Gogh's "Starry Night," and having an AI guide recount not just the history of the piece, but also how it connects with other works you've shown interest in.

On the technical side, AI contributes to the realism of VR experiences through various means. This includes optimizing 3D environments, enhancing graphical details in real-time, and even simulating complex physical interactions. AI models can generate textures, landscapes, and even entire worlds that evolve dynamically, offering infinite possibilities for exploration and interaction.

In virtual shopping, AI-driven VR experiences are changing the landscape of retail. Shoppers can now enter a virtual store, personally styled to their preferences, and even receive AI-assisted recommendations. The virtual try-on experiences, where AI helps render clothing or accessories realistically on your avatar, further enhances the convenience and realism of online shopping.

Another noteworthy application is in the therapeutic domain. Therapists are employing AI-powered VR to help treat conditions such as PTSD (Post-Traumatic Stress Disorder), anxiety, and phobias. These systems can create controlled environments where patients are gradually exposed to their fears or triggers. AI tailors these experiences to individual's progress, ensuring a safe and effective therapeutic process.

Moreover, AI in VR is making fitness routines more engaging and effective. Envision a virtual personal trainer who not only guides you through exercises but also adjusts your workout in real-time based on your performance and physiological data. This level of customization was unimaginable just a few years ago but is now increasingly becoming a reality.

Nonetheless, the integration of AI in VR isn't without challenges. Privacy concerns, data security, and ethical considerations are paramount as these technologies increasingly interact with our personal spaces and data. Developers and policymakers must collaborate to create frameworks that address these concerns while fostering innovation and growth.

Looking ahead, the future of AI-enhanced VR experiences seems nothing short of revolutionary. As AI algorithms become more sophisticated and VR technology continues to advance, the boundary between the virtual and the real world will blur even further. This convergence opens up unprecedented opportunities for creativity, education, and interaction.

Furthermore, the concept of the "metaverse"—a collective virtual shared space—might soon evolve from science fiction to commonplace reality. AI will be the backbone of this digital universe, intelligently managing interactions, environments, and even economies within these virtual realms. The implications are vast and extend into how we work, socialize, and entertain ourselves.

In essence, AI is the catalyst propelling VR into new heights of interactivity, realism, and personalization. As we stand on the cusp of this transformative era, the potential for enriching our lives through AI-enhanced VR experiences is immense. Whether for entertainment, education, therapy, or social interaction, AI in Virtual Reality is poised to redefine how we perceive and interact with the digital world.

In conclusion, the fusion of AI with Virtual Reality isn't just enhancing our experiences; it's fundamentally changing the way we perceive reality itself. The possibilities are limited only by our imagination, making this an incredibly exciting field to watch as it evolves. The future promises VR experiences that are not just more realistic, but more intimately tailored to who we are, what we need, and how we interact with the world around us.

CHAPTER 7:
AI IN COMMUNICATION

A dvancements in AI are revolutionizing the way we communicate, transforming interactions in ways previously unimaginable. By incorporating natural language processing (NLP), AI systems can now understand and generate human language with startling accuracy. Businesses are leveraging AI to enhance customer service, deploying chatbots and virtual assistants that can handle inquiries round-the-clock, offering personalized and immediate assistance. AI-powered smart messaging apps streamline communication, automatically suggesting context-aware replies, translating languages in real-time, and even detecting the emotional tone of the conversation. These innovations not only bolster efficiency but also foster more meaningful and accessible human connections, highlighting AI's immense potential to enrich our daily lives.

Natural Language Processing

When it comes to communication, Natural Language Processing (NLP) stands as one of the most transformative technologies in AI. NLP ref ers to the ability of machines to understand, interpret, and generate human language. By leveraging NLP, various applications have been developed that can enhance our daily communications, making them more efficient and impactful.

One of the primary roles of NLP is in the realm of text-based interactions. This includes anything from simple chatbots to complex

AI-driven customer service platforms. Imagine contacting a company for support and having your issues resolved instantly by an intelligent agent capable of understanding and responding to your queries without human intervention. This is the power of NLP in action, streamlining customer experiences and freeing up human agents for more complex tasks.

NLP also revolutionizes the way we write and edit documents. Think about intelligent writing assistants, which not only correct grammatical errors but also suggest improvements in tone, style, and readability. These tools go beyond simple corrections, offering insights that can make even the most mundane reports engaging and easier to read. They can localize language, adjust the complexity based on the target audience, and even ensure that the content aligns with brand guidelines.

Another valuable application of NLP is in real-time language translation. In our increasingly globalized world, the ability to communicate across languages effortlessly is invaluable. Services providing instant translation enable people from different linguistic backgrounds to converse fluidly, breaking down barriers and fostering greater understanding and collaboration. Whether it's translating a menu while dining abroad or facilitating a business meeting between multinational partners, NLP-powered translation systems can make communication seamless.

NLP has made significant strides in sentiment analysis, enhancing how we interpret opinions and emotions expressed in text. Companies use sentiment analysis to gauge the public perception of their brands, campaigns, and products. By analyzing social media posts, reviews, and feedback, businesses can derive actionable insights that help them make more informed decisions. Understanding the emotional tone behind text can also enhance customer relations, enabling more tailored and empathetic responses.

Voice recognition and virtual assistants are perhaps the most ubiquitous examples of NLP at work in our daily lives. Siri, Alexa, and Google Assistant are not just virtual helpers—they are conversational agents that understand and respond to voice commands. These assistants perform a myriad of tasks, from setting reminders and playing music to controlling smart home devices. Their ability to understand natural language queries and provide relevant, context-aware responses highlights the sophistication of modern NLP algorithms.

NLP isn't just about understanding human language; it's also about generating it. AI-driven content generation tools can create coherent and contextually relevant articles, reports, and even creative writing pieces. This capability is particularly revolutionary in fields such as journalism, marketing, and entertainment. For instance, news agencies use AI to produce quick news summaries, while marketers employ it to generate engaging content tailored to specific audiences.

Consider the advancements in conversational agents designed for healthcare settings. NLP-powered systems like chatbots and virtual health assistants can triage patients, provide medical information, and even offer emotional support. By automating these interactions, healthcare providers can extend their reach and improve patient care while allowing human professionals to focus on more critical aspects of treatment.

NLP also benefits educational environments. Intelligent tutoring systems can adapt to a student's learning style and pace, offering personalized educational content. These systems understand and respond to the student's queries, providing explanations and feedback that align with their specific needs. Additionally, NLP can assist in grading, offering educators more time to dedicate to teaching rather than administrative tasks.

In the realm of business intelligence, NLP enables the extraction of valuable insights from vast datasets of unstructured text. By analyzing documents, emails, and reports, AI systems can summarize information, identify trends, and even predict future outcomes. This capability is crucial for decision-makers who need to process large amounts of information swiftly and accurately.

As AI continues to evolve, so does the scope of NLP. Future advancements may bring even more nuanced understanding and generation of human language, enabling more natural and meaningful interactions between humans and machines. This evolution will undoubtedly open new avenues for enhancing communication across various domains, further integrating AI into our everyday lives.

Natural Language Processing stands at the frontier of AI innovation, significantly impacting how we communicate, access information, and interact with technology. Its applications are vast and varied, touching every aspect of our daily lives from personal assistant tools to global business solutions. As NLP technology becomes more sophisticated, its ability to simplify and enhance communication will undoubtedly lead to a future where seamless interaction with machines is not just a convenience but a cornerstone of our everyday experiences.

In the end, NLP is about bridging the gap between human intent and machine understanding, creating a symbiotic relationship where both can thrive. By continuing to advance this technology, we move closer to a world where AI not only supports but also elevates the quality and efficiency of our communication, making our day-to-day interactions richer and more fulfilling.

AI in Customer Service

As customer expectations continue to skyrocket, businesses are under constant pressure to provide faster, more efficient services. This is where AI in customer service shines, revolutionizing how companies interact with their clients. From chatbots to predictive analytics, AI technologies have embedded themselves deeply into the customer service industry, creating a landscape where responses are quicker, more accurate, and highly personalized.

The incorporation of AI in customer service starts with chatbots, the most recognizable form of this technology. These automated conversational agents can answer queries, direct users to relevant resources, and even complete transactions without human intervention. Equipped with Natural Language Processing (NLP) capabilities, modern chatbots understand and interpret user inputs in various languages and dialects, offering a seamless interaction. They operate 24/7, providing instant responses and ensuring that customer grievances are addressed immediately. This not only enhances user satisfaction but significantly reduces operational costs.

Beyond chatbots, AI can delve deeper into customer interactions through sentiment analysis. By analyzing the tone and language of customer communications—whether it's through emails, chat logs, or social media posts—AI can gauge the customer's emotions and intentions. This allows companies to address negative feedback proactively and tailor their responses more effectively. It's a powerful tool for maintaining brand loyalty and improving overall customer experience.

Moreover, AI-driven predictive analytics is another game-changer in customer service. By sifting through mountains of customer data, AI can identify patterns and predict future behaviors. For instance, AI can forecast when a customer is likely to need assistance based on their purchase history and browsing patterns. This enables businesses to offer preemptive support, reaching out to clients before they even encounter an issue. Predictive analytics also help in upselling and cross-selling products by suggesting items that a particular customer might be inclined to buy, making the interaction both useful and persuasive.

AI doesn't just improve the front-end customer experience; it also optimizes the backend processes that support these interactions. Automated workflows and intelligent routing systems ensure that customer queries are directed to the most appropriate departments or agents, reducing wait times and the chances of miscommunication. For instance, AI can automatically route a technical query to a tech support team while directing billing issues to the finance department. This efficient handling of queries means that customers receive expert help faster, enhancing their overall experience.

In addition, voice recognition and AI-powered IVR (Interactive Voice Response) systems are revolutionizing call centers. AI can accurately interpret voice commands and direct calls to the right department or provide automated responses to common queries. This technology not only shortens call times but also frees up human agents to handle more complex and nuanced customer issues requiring a human touch.

Further elevating customer service, AI-powered Customer Relationship Management (CRM) systems collect and analyze customer data in real time to offer a complete view of the customer journey. These systems can recommend the best course of action during an interaction, whether it's suggesting useful content, providing insights into past interactions, or recommending products. Enhanced CRM systems contribute to a more personalized customer experience by ensuring that every touchpoint is informed and contextual, thereby increasing the likelihood of a satisfactory resolution.

Employee training and support are other areas where AI is making a significant impact. AI-driven training modules can simulate various customer interactions, allowing customer service representatives to practice and refine their skills in a controlled environment. These training programs can adapt in real-time based on the employee's performance, providing tailored recommendations for improvement. Additionally, AI tools can assist agents during live customer interactions by suggesting responses and offering solutions based on previous similar cases.

Personalized AI assistants for customer service agents are also becoming indispensable. These assistants provide real-time support, helping agents to retrieve information quickly, suggest next best actions, and even create summary reports post-interaction. This helps to ensure that the agents can focus more on the emotional and relational aspects of customer service, areas where the human touch is irreplaceable.

While the benefits of AI in customer service are extensive, it is essential to address the challenges. Bias in AI algorithms can lead to unfair treatment of certain customer groups if not carefully monitored and rectified. Moreover, the integration of AI into customer service workflows requires a paradigm shift in how companies think about customer interactions. Companies need to ensure that their AI systems are transparent and that humans retain oversight of AI-driven processes to prevent any unintended consequences.

The future of AI in customer service holds even more promise with advancements in machine learning and AI-human collaboration. As these technologies evolve, we are likely to see even more sophisticated forms of customer interaction. For instance, AI could provide real-time translations in multilingual support, making it much easier for companies to operate globally and offer support in multiple languages seamlessly.

Moreover, as AI systems become better at understanding context and nuances in human communication, the scope of what can be automated will expand. We may reach a point where the line between human and AI interaction becomes almost imperceptible, with AI taking over mundane tasks entirely and allowing human agents to focus solely on complex issues that necessitate human empathy and creativity.

In closing, AI in customer service is not just a fleeting trend but a fundamental shift in how businesses approach customer interactions. By automating routine queries, enhancing personalization, and improving operational efficiency, AI is setting new standards in customer satisfaction. The journey towards fully integrating AI in customer service is just beginning, but its influence is already transforming the landscape. As we continue to embrace and refine these technologies, the potential for enhanced customer experiences stands as a beacon for what's possible when human ingenuity meets technological innovation.

Smart Messaging Apps

In the orbit of modern communication, smart messaging apps have emerged as powerful tools, fundamentally transforming how we interact in our day-to-day lives. These applications harness the potential of artificial intelligence to enhance user experience, making conversations seamless, efficient, and engaging. But what sets them apart from traditional messaging apps? The answer lies in their intelligent features, which go far beyond simple text exchange.

At the heart of smart messaging apps are advanced machine learning algorithms and natural language processing (NLP) techniques. These technologies enable apps to understand, interpret, and respond to human language in a way that feels eerily natural. For instance, predictive text and auto-complete features are now sophisticated enough to suggest not just words, but entire phrases and sentences, based on the context of the conversation. This doesn't just save time; it also enhances the flow of dialogue by reducing the cognitive load on the user.

Beyond mere text suggestions, smart messaging apps are adept at sentiment analysis. These apps can gauge the emotional tone of a conversation and provide responses that align with that emotion. Imagine expressing frustration in a message and receiving a reply that shows empathy or offers a solution. The emotional intelligence embedded in these apps can make digital conversations feel more personal and considerate.

One of the most captivating features is the integration of intelligent chatbots. These AI-powered bots can handle a myriad of tasks: answering queries, setting appointments, conducting surveys, and even performing customer service functions. By automating these tasks, smart messaging apps free up time for users, making interactions not just smarter, but also more productive. These chatbots continually learn from interactions, becoming more sophisticated and capable of handling complex queries over time.

Smart messaging apps also leverage AI to enhance multimedia sharing. AI can automatically tag photos, suggest relevant emojis, stickers, or GIFs, and even generate captions. Visual content often speaks louder than words, and smart messaging apps ensure your multimedia communication is as intuitive and expressive as possible. By predicting and suggesting the right visual aids, these apps augment the traditional text-based experience, making it richer and more engaging.

In professional settings, smart messaging apps are proving invaluable for boosting productivity. They integrate seamlessly with other AI-driven tools such as calendars, project management software, and virtual assistants. Imagine discussing a project deadline in a chat and having the app automatically update your calendar or send reminders to team members. It's a nuanced form of automation that keeps everyone in sync without the need for constant manual updates.

Security is another pivotal aspect where AI plays a significant role in smart messaging apps. Advanced encryption methods and AI-based threat detection safeguard user data and ensure privacy. By analyzing patterns and identifying potential anomalies, AI can preemptively block phishing attempts and other malicious activities. This creates a secure environment where users can communicate freely without worrying about data breaches.

Furthermore, smart messaging apps are incredibly adept at language translation. Leveraging machine translation technologies like neural networks, these apps can provide instant translations within chat windows. This breaks down language barriers and fosters seamless communication across different linguistic backgrounds. Whether you're texting a friend in another country or conducting international business, language is no longer a barrier but a bridge.

The impact of smart messaging apps extends to the world of commerce as well. Businesses are increasingly using these apps to engage with customers in real-time, provide personalized recommendations, and even complete transactions within the chat window. AI-driven customer insights help businesses understand user behavior and preferences, allowing for highly targeted interactions. This level of personalization enhances customer satisfaction and loyalty.

Smart messaging apps are also integrating augmented reality (AR) features, which add a new dimension to digital communication. Users can now overlay digital elements onto their real-world environment, making video calls more interactive and fun. Whether it's trying out virtual makeup or placing virtual furniture in your living room, the fusion of AI and AR creates an immersive experience that traditional messaging platforms can't match.

The evolution of smart messaging apps is still in its nascent stages, with new innovations on the horizon. Emerging trends such as the integration of voice assistants and the rise of conversational interfaces promise to further revolutionize how we communicate. As AI continues to advance, we can expect even more intelligent, intuitive, and empowering features that will make smart messaging apps indispensable in our daily lives.

In conclusion, smart messaging apps are much more than just tools for communication. They are becoming comprehensive platforms that combine the best of AI to offer a seamless, secure, and highly personalized user experience. By making our interactions more efficient and engaging, these apps hold the promise of not only simplifying communication but also enriching our lives in ways we're just beginning to explore.

CHAPTER 8:
AI IN TRANSPORTATION

A I in transportation is revolutionizing how we move by enhancing efficiency, safety, and convenience. From autonomous vehicles that promise to reduce accidents and ease urban congestion to intelligent traffic management systems that optimize routing and minimize delays, AI's impact is profound. Moreover, ride-sharing services are leveraging AI to create more efficient, on-demand transportation models, reducing both costs and environmental impact. Through advanced machine learning algorithms, these systems continuously improve their decision-making processes, making smart transportation a reality. As we embrace these innovations, the potential for AI in transportation extends far beyond our current horizons, paving the way for a future where mobility is seamless, sustainable, and accessible to all.

Autonomous Vehicles

Autonomous vehicles, often referred to as self-driving cars, have captured public interest and imagination. At their core, these vehicles utilize Artificial Intelligence (AI) to navigate and operate without human intervention. Their impact on transportation could be revolutionary, transforming the way we think about travel, logistics, and even urban planning.

The magic behind autonomous vehicles lies in a blend of sensor technology, complex algorithms, and machine learning. Equipped with a multitude of sensors—including cameras, radar, and LIDAR systems—these vehicles constantly scan their surroundings, creating a detailed, real-time map. This data is then analyzed by sophisticated AI algorithms that make split-second decisions, enabling the vehicle to react to changing road conditions, avoid obstacles, and safely transport passengers to their destinations.

One of the most promising aspects of autonomous vehicles is their potential to significantly reduce traffic accidents. According to the National Highway Traffic Safety Administration (NHTSA), human error accounts for 94% of serious crashes. By removing the human element, which is prone to error and distraction, autonomous vehicles could dramatically enhance road safety. Imagine a world where drunk driving, texting while driving, and even basic driver fatigue are no longer factors in traffic fatalities.

Another remarkable benefit is the potential to improve traffic flow and reduce congestion. Autonomous vehicles can communicate with each other and with traffic management systems, optimizing routes and speed to minimize traffic jams. Picture a cityscape where cars move seamlessly, traffic lights are almost obsolete, and commute times are slashed. This would not only save time but also reduce fuel consumption and emissions, contributing to a greener environment.

Accessibility is another area where autonomous vehicles can make a transformative impact. Elderly individuals, people with disabilities, or those unable to drive for any reason would gain newfound independence, improving their quality of life. An autonomous vehicle could be summoned with a simple app, providing door-to-door service and integrating seamlessly with other forms of public transport.

The potential applications extend beyond personal travel. Autonomous trucks and delivery vehicles could revolutionize logistics and supply chains. Companies like UPS and Amazon are already testing self-driving delivery vans and drones. These innovations promise to expedite delivery times, reduce operational costs, and mitigate the risks associated with long-haul trucking, such as driver fatigue. It's not hard to envision a near-future where your online order arrives at your doorstep much faster, thanks to a network of autonomous delivery systems.

However, the road to widespread adoption of autonomous vehicles is not without challenges. One of the primary concerns is the complex regulatory landscape. Both federal and state governments need to establish comprehensive regulations that ensure safety while fostering innovation. This requires collaboration between policymakers, tech companies, and the automotive industry.

Public trust is another significant hurdle. While tech enthusiasts may be ready to embrace self-driving cars, a broader segment of the population remains skeptical. Incidents involving autonomous vehicles during testing phases have been widely publicized, fueling concerns about their reliability. To build public confidence, rigorous testing and transparent reporting are essential. Companies must demonstrate that their autonomous systems are not only as safe as human drivers but significantly safer.

Cybersecurity is also a critical issue. As vehicles become more connected, they are at risk of being targeted by hackers. Ensuring the security of these systems is paramount to prevent potentially catastrophic scenarios. Robust encryption, regular software updates, and fail-safe mechanisms are essential components of a secure autonomous vehicle network.

Ethical considerations cannot be ignored either. Autonomous vehicles will need to make complex decisions in scenarios where harm is unavoidable. This raises questions about the moral framework within which these AI systems operate. Who is liable in the event of an accident? How should a vehicle prioritize the lives of its passengers versus pedestrians? These are complex issues that require thoughtful deliberation and input from ethicists, engineers, and legal experts.

Despite these challenges, major players in the automotive and tech industries are making significant strides. Companies like Tesla, Google's Waymo, and Uber are at the forefront of autonomous vehicle development, investing billions in research and testing. Their progress is bringing us ever closer to a future where self-driving cars are a mainstream reality.

In terms of urban planning, the implications of autonomous vehicles are vast. Parking structures and spaces could be radically redefined when cars are capable of dropping passengers off and then driving themselves to optimal parking locations. Urban planners could reclaim significant amounts of real estate currently devoted to parking, opening up spaces for parks, residential areas, and community centers.

While fully autonomous vehicles may still be several years away from becoming a common sight on our roads, the incremental advancements being made today are setting the stage for widespread adoption. Features like adaptive cruise control, automatic emergency braking, and lane-keeping assistance are becoming standard in many new cars, serving as building blocks toward fully autonomous systems.

Insurance is another area undergoing transformation. As the risk factors associated with human drivers diminish, insurance models will need to adapt. Policies may shift from personal liability to product liability, with manufacturers shouldering more responsibility. This shift could also lead to lower insurance premiums for consumers, making driving more affordable.

Moreover, the integration of autonomous vehicles into the broader transportation ecosystem could spur advances in other areas. For example, smart infrastructure, such as intelligent traffic lights and advanced road sensors, can work in tandem with autonomous vehicles to create a highly efficient transportation network. This synergy could pave the way for smart cities, where technology is seamlessly integrated into daily life to enhance urban living.

In conclusion, autonomous vehicles represent a profound leap forward in the realm of transportation, driven by the power of AI. They promise to enhance safety, improve efficiency, and provide new levels of accessibility, all while reshaping our urban landscapes. The journey to full autonomy will undoubtedly be complex and challenging, but the potential rewards are immense. As technology continues to advance, the vision of a world where autonomous vehicles are the norm, rather than the exception, draws ever closer, heralding a new era in human mobility.

Intelligent Traffic Management

When we talk about the transformative potential of artificial intelligence in transportation, intelligent traffic management stands out as one of the most revolutionary applications. AI offers unprecedented capabilities to optimize the flow of vehicles on our roads, reducing congestion, improving safety, and enhancing the overall efficiency of our transportation networks. Let's delve into how AI is reshaping the future of traffic management.

Imagine a world where traffic jams are a thing of the past. This is not a far-fetched dream but a rapidly approaching reality thanks to AI-driven traffic management systems. These systems utilize real-time data from various sources, such as cameras, sensors, and even smartphones, to monitor traffic conditions and make instantaneous decisions. By analyzing this data, AI algorithms can predict traffic patterns, adjust traffic signals, and reroute drivers to reduce congestion and emissions.

One key technology in intelligent traffic management is adaptive signal control. Traditional traffic lights operate on fixed schedules, which can be inefficient during fluctuating traffic volumes. Adaptive signal control systems, however, use machine learning algorithms to dynamically adjust signal timings based on real-time traffic conditions.

This level of responsiveness can significantly improve traffic flow, reduce waiting times at intersections, and decrease the likelihood of accidents.

Let's consider an example. In Pittsburgh, Pennsylvania, AI-driven adaptive traffic signals have been successfully implemented. These smart signals have led to a 40% reduction in traffic wait times and a 26% decrease in travel time through congested areas. Such technology not only enhances the commuter experience but also contributes to a reduction in fuel consumption and greenhouse gas emissions, dovetailing nicely with broader sustainability goals.

In addition to adaptive signal control, AI-powered traffic management systems often incorporate predictive analytics. Predictive models can forecast traffic conditions several hours or even days in advance by analyzing historical data, weather conditions, special events, and other variables. This allows transportation authorities to proactively manage traffic flow, implement control measures, and issue warnings to drivers about potential delays or hazards.

Another exciting development is the integration of AI with vehicle-to-everything (V2X) communication technologies. V2X enables vehicles to communicate with each other and with infrastructure, such as traffic lights and road signs. When combined with AI, this communication can facilitate coordinated vehicle movements, effectively creating a harmonious traffic ecosystem. For instance, AI can manage the speed and flow of vehicular traffic entering a highway, reducing the stop-and-go waves that often lead to congestion. In turn, this synergy can elevate the overall efficiency and safety of our roadways.

Public transportation systems also benefit from AI-driven traffic management. Buses and trams can be given priority at traffic signals, ensuring more punctual services. AI can also optimize routes based on passenger demand, leading to better resource utilization and more efficient public transit operations. For instance, in the city of Barcelona, smart traffic lights prioritize buses and emergency vehicles, significantly improving their travel times and reliability.

The safety implications of intelligent traffic management are profound. AI can identify and predict high-risk situations, such as potential collisions at intersections or sudden braking events. By alerting drivers in real-time, these systems can prevent accidents and save lives. Moreover, AI-based traffic monitoring can aid in swift incident detection and management, expediting emergency response efforts.

Intelligent traffic management is also opening up new avenues for data collection and analysis. Cities equipped with AI-driven systems can gather vast amounts of traffic-related data, which can be harnessed for various urban planning purposes. This data allows city planners to better understand traffic patterns, assess the effectiveness of infrastructure projects, and make data-driven decisions to address transportation challenges.

It is not just urban areas that are set to benefit from intelligent traffic management. Rural and suburban regions, often facing different kinds of traffic issues, can also leverage AI solutions. For example, AI can help in optimizing the timing and placement of road maintenance operations, leading to minimal disruption and enhanced road safety. Similarly, traffic flow on rural highways can be improved by AI-enabled dynamic signage that provides real-time information to drivers.

The funding and implementation of intelligent traffic management systems are significant considerations. While the initial costs can be high, the long-term benefits in terms of reduced congestion, lower emissions, and enhanced safety make it a worthwhile investment. Public-private partnerships could play a crucial role in accelerating the deployment of these advanced systems, ensuring that cities and regions can reap the benefits sooner rather than later.

As we look to the future, the synergy between autonomous vehicles and intelligent traffic management presents an exciting frontier. Autonomous vehicles, integrated with smart traffic systems, have the potential to transform our transportation landscape radically. They can communicate with traffic infrastructure to optimize their routes, reduce travel times, and enhance safety. Autonomous vehicles and intelligent traffic management systems can form a cohesive, self-regulating ecosystem that maximizes efficiency and minimizes human error.

Of course, the adoption of intelligent traffic management is not without its challenges. Privacy concerns regarding the vast amounts of data collected, the cost of upgrading outdated infrastructure, and the necessity for standardized protocols and regulations are significant hurdles. However, with the socio-economic and environmental benefits that intelligent traffic management promises, addressing these challenges is a crucial step towards a smarter, more efficient future.

In conclusion, AI-driven intelligent traffic management holds the key to revolutionizing how we experience transportation. By optimizing traffic flow, reducing congestion, enhancing safety, and facilitating more sustainable urban environments, AI presents an opportunity to create smarter, more liveable cities. The future of transportation is not just about moving from point A to point B; it's about doing so in the most efficient, safe, and sustainable manner possible. Intelligent traffic management, underpinned by the power of AI, is set to drive this transformation forward, promising a brighter future on our roads.

Ride-Sharing Services

When we talk about AI in transportation, one of the most transformative areas is ride-sharing services. These platforms have evolved rapidly, leveraging AI to shift paradigms in how we commute. Companies like Uber, Lyft, and Didi Chuxing have integrated artificial intelligence to optimize routes, match riders with drivers, provide safety measures, and even predict user demand. The result? A more seamless, efficient, and convenient ride-sharing experience.

AI helps ride-sharing services by making dynamic pricing possible. Dynamic pricing, or surge pricing, ensures that the number of available drivers meets the rider demand, especially during peak times. It can analyze patterns and anticipate demand spikes by considering factors like weather, special events, and time of day. This system not only balances supply and demand but also incentivizes drivers to be available during high-demand periods, ensuring that riders don't face long wait times.

Moreover, AI fundamentally enhances ride-matching. Instead of randomly pairing riders and drivers, algorithms assess a multitude of variables including distance, traffic conditions, and driver ratings. These criteria help ensure the most efficient and satisfactory matches, thereby reducing wait times and ensuring a quicker route to the destination. The endgame? A more satisfying experience for both the driver and the rider.

Safety is another critical aspect where AI steps in. From facial recognition to verify a driver's identity before they start their shift, to real-time monitoring systems that can detect unusual driving patterns suggesting fatigue or intoxication, AI adds layers of security previously unimaginable. These technologies give both drivers and riders an added sense of safety, fostering trust in the ride-sharing ecosystem.

In addition, predictive maintenance empowered by AI ensures fleet reliability. Monitoring vehicle health in real-time allows early detection of potential issues, like engine failure or brake wear, before they become serious problems. This preemptive approach not only keeps vehicles roadworthy but also enhances rider safety. AI algorithms can analyze data from thousands of cars to predict which ones are likely to need maintenance soon, scheduling service appointments more efficiently than ever before.

The impact of AI in ride-sharing is not just limited to operational mechanics. User experience is profoundly affected as well. AI-driven chatbots provide immediate customer support, capable of handling everything from cancellations and refunds to lost items and complaints. These chatbots offer 24/7 assistance, ensuring issues are resolved promptly without requiring human intervention.

Socially, the implications are vast. Ride-sharing apps using AI have democratized transportation in many urban areas, making it more accessible to those who do not own vehicles. This democratization also extends to sustainability efforts. By optimizing routes and encouraging carpooling, AI helps reduce the number of vehicles on the road, contributing to lower emissions and less traffic congestion.

With AI's capability to understand and analyze user preferences, personalized recommendations become another fascinating aspect. Whether it's suggesting the best times to travel or recommending certain types of rides based on past behavior, AI tailors each user experience to meet individual preferences. Frequent users may receive loyalty rewards or unique offers, making the service not just a utility but a personalized experience.

Data analytics plays a massive role in revolutionizing this space. By analyzing historical data, AI systems can provide insights into traffic patterns, enabling cities to make informed decisions on infrastructure planning and traffic management. Imagine a world where city planners and ride-sharing companies work in tandem, using data to minimize congestion and improve overall urban mobility.

Of course, we're not without challenges and controversies. Concerns about data privacy and algorithmic fairness continue to loom large. Algorithms that manage surge pricing or rider-driver matching need to be transparent and fair. It's crucial that these systems do not inadvertently discriminate against certain demographics or neighborhoods. Addressing these issues head-on, companies are increasingly focusing on ethical AI practices to ensure inclusive and impartial services.

Looking ahead, the integration of AI with other emerging technologies like autonomous vehicles represents the next frontier in ride-sharing. Though fully autonomous ride-sharing fleets are not a reality yet, companies are relentlessly working toward this goal. Imagine summoning an autonomous vehicle via your ride-sharing app – a car that arrives without a driver, takes you to your destination, and then continues to its next passenger efficiently and responsibly. Merging AI with autonomous driving technology promises to redefine what we know about private, on-demand transportation.

Ultimately, the marriage of AI and ride-sharing is a testament to how technology can transform daily activities in ways both minor and monumental. From the minute adjustments in pricing and routing to the overarching goals of safety and sustainability, AI proves to be an indispensable component of modern transportation. The convenience it brings to our lives suggests an inspiring future, one where each journey—no matter how small—is optimized, secure, and tailored to our needs.

CHAPTER 9:
AI IN RETAIL AND SHOPPING

The evolution of retail and shopping through the integration of AI isn't just a futuristic concept—it's already reshaping the way we experience commerce today. Imagine walking into a store where your preferences are already known, thanks to AI-driven personalized shopping experiences that cater to your unique tastes and needs. This transformation doesn't end with the consumer; behind the scenes, AI in supply chain management is streamlining operations, predicting stock needs, and ensuring timely delivery of products with smart warehouses that operate with remarkable efficiency. From the moment products are sourced to when they land on the shelves, AI's rapid data processing and predictive capabilities are enhancing every step, making retail not only more efficient but also infinitely more responsive to consumer demands.

Personalized Shopping Experiences

Imagine walking into your favorite clothing store, and instead of spending hours sifting through racks, a helpful assistant immediately guides you to items that align perfectly with your tastes. Now, picture that assistant knowing your size, fabric preferences, and even your favorite colors. This isn't some far-off fantasy—it's the promise of AI-driven personalized shopping experiences that are transforming the retail sector.

Artificial Intelligence is rapidly revolutionizing how we shop, providing a highly individualized experience that was previously unimaginable. Retailers are now using AI to analyze customer data, understand buying behaviors, and predict future purchases. These insights are then used to create a shopping journey that feels uniquely tailored to each individual. A visit to an online store becomes a seamless experience, mirroring your needs and desires almost as if the website could read your mind.

One of the most straightforward but impactful uses of AI in retail is through recommendation systems. Algorithms analyze your past purchases, browsing history, and even social media activity to suggest products that you are most likely to buy. Think of it as a digital concierge, refining its suggestions the more it gets to know you. This feature has become a staple on e-commerce platforms, where personalized recommendations can significantly boost sales and enhance customer satisfaction.

But AI is not just relegated to the online space. Physical stores are also leveraging this technology to offer a more customized shopping experience. In some advanced retail settings, AI-driven bots can greet you at the entrance and offer tailored recommendations. These bots can even adjust their suggestions based on real-time inputs, such as items you show interest in during your visit. The integration of AI with augmented reality (AR) further elevates this experience. For example, beauty retailers are using AR mirrors that allow customers to virtually try on makeup, complete with AI-driven recommendations on shades and products that best suit their skin tone.

Furthermore, AI is capable of revolutionizing the way we receive discounts and promotions. Instead of generic, one-size-fits-all promotions, retailers can now customize offers based on individual shopping habits and preferences. Imagine receiving a discount code for your favorite brand just days before you intend to purchase a product. This level of personalization makes customers feel valued and appreciated, fostering loyalty and repeat business.

Personalized shopping experiences also extend to customer service. Traditional customer support methods often involve a significant time commitment, waiting on hold, and explaining your issue multiple times. AI, however, allows for a more streamlined interaction. Chatbots and virtual assistants can provide immediate responses to common queries, guide you through the purchasing process, and even troubleshoot issues post-purchase. Advanced AI-driven support systems can learn from past interactions to provide even more accurate and friendly service over time.

The magic behind these personalized interactions lies in data—lots of it. Retailers collect data from various touchpoints such as purchase history, browsing patterns, and even social media activity. AI algorithms analyze this data to uncover patterns and insights that are invisible to the human eye. By continually learning from this data, these systems can adapt and refine the shopping experience, making it more intuitive and enjoyable with each interaction.

This hyper-personalization, while amazing, raises essential questions about privacy and data security. Consumers need to trust that their information is being used responsibly and stored securely. Retailers must ensure compliance with data protection laws and prioritize transparency, giving customers control over their data. Building this trust is crucial for the long-term success of AI in retail and requires a balanced approach that safeguards customer privacy while delivering personalized experiences.

AI's ability to create personalized shopping experiences isn't solely for consumer benefit; it significantly aids retailers in inventory management, demand forecasting, and streamlined logistics. Knowing what customers want allows retailers to stock items more efficiently, reducing waste and increasing profitability. Predictive analytics can anticipate trends and shifts in consumer preferences, allowing retailers to stay ahead of the curve and meet demand more effectively.

Looking to the future, the potential for further innovation in personalized shopping experiences is boundless. Emerging technologies such as deep learning, natural language processing, and computer vision will likely play pivotal roles. For instance, imagine AI systems capable of understanding and reacting to your emotional state while shopping, offering different products and services based on whether you're feeling happy, stressed, or excited.

As personalization becomes more advanced, consumers will start to expect this level of service across all retail interactions. The challenge for retailers will be to integrate these advanced AI systems seamlessly into their operations, ensuring that the technology enhances rather than detracts from the shopping experience. The ultimate goal is to create a shopping environment where the journey is as enjoyable as the destination, making every interaction feel like it was crafted just for you.

In an era where consumer expectations are higher than ever, leveraging AI to deliver personalized shopping experiences is no longer a luxury but a necessity. Retailers who embrace this technology will find themselves at the forefront of a customer-first revolution, equipped to meet the needs of the modern shopper in ways that were once considered the stuff of science fiction. As AI continues to evolve, so too will our shopping experiences, becoming more intuitive, enjoyable, and uniquely tailored to our individual tastes and preferences.

AI in Supply Chain Management

As the retail and shopping landscape continues to evolve, Artificial Intelligence (AI) has cemented its role as a transformative force in supply chain management. Whether we're talking about forecasting demand, reducing operational costs, or enhancing customer experiences,

AI is making supply chains smarter and more efficient. The complexity of the modern retail supply chain, which spans multiple geographies and intricate networks, demands sophisticated solutions that can adapt in real-time. That's where AI steps in, providing capabilities that were once the stuff of science fiction.

One significant advantage of AI in supply chain management is its ability to forecast demand more accurately. Traditional methods relied heavily on historical sales data and could be painfully slow and inaccurate, leading to either stockouts or overstock situations. AI algorithms, on the other hand, can analyze massive amounts of data from various sources—including social media trends, weather forecasts, and economic indicators—to provide a much more precise prediction of future demand. This increased accuracy allows retailers to keep the right amount of stock, reducing both lost sales and excess inventory carrying costs.

Moreover, AI facilitates intelligent inventory management. By leveraging machine learning and predictive analytics, AI-driven systems can optimize stock levels at various points in the supply chain. These systems take into account factors such as seasonal changes, promotional events, and even competitor actions to ensure that products are available when and where they are needed. This kind of precision not only enhances customer satisfaction but also minimizes waste and boosts profitability.

But AI's role in supply chain management doesn't stop at inventory and demand forecasting. Route optimization is another crucial area where AI is making a significant impact. Traditional route planning can be time-consuming and often fails to consider real-time variables such as traffic conditions, weather disruptions, and unexpected delays. AI-driven platforms can process real-time data to suggest the most efficient routes for logistics fleets. These optimizations can lead to faster delivery times, reduced fuel consumption, and lower operational costs—all thanks to AI.

Additionally, AI can improve supplier relationship management. By analyzing performance metrics and compliance data, AI systems can identify the most reliable and cost-effective suppliers. They can also predict potential disruptions in the supply chain and suggest alternative suppliers, ensuring that operations run smoothly even when complications arise. This predictive capability is invaluable in maintaining the robustness of supply chains and helps retail businesses avoid costly interruptions.

Another game-changing aspect of AI in supply chain management is its role in enhancing supply chain transparency. Blockchain technology, combined with AI, allows for unparalleled traceability of goods from raw material to finished product. This transparency is particularly vital for sectors such as food and pharmaceuticals, where knowing the provenance of products is crucial for safety and regulatory compliance. AI can quickly identify anomalies in the supply chain, such as delayed shipments or deviations from standard procedures, allowing for rapid corrective action.

AI also aids in automating procurement processes. Traditional procurement required human oversight to review purchase orders, verify invoices, and manage supplier contracts. AI-powered systems can automate these tasks, freeing up human resources for more strategic activities. These systems can also analyze vast datasets to identify cost-saving opportunities, such as bulk purchasing or identifying the optimal times to buy specific items.

Machine learning algorithms play a pivotal role in quality control within the supply chain. Automated inspection systems can quickly analyze products using image recognition and other sensory data to identify defects or inconsistencies. This capability ensures that only high-quality products make it to the retail shelves, reducing returns and enhancing brand reputation.

Furthermore, AI-driven supply chain management systems can contribute to enhanced sustainability. By optimizing routes and loads, reducing waste, and improving demand forecasting, AI can help retail businesses decrease their carbon footprint. Sustainability is becoming an ever-more crucial factor in consumer choices, and companies that can demonstrate their commitment to eco-friendly practices stand to win greater customer loyalty.

In the realm of dynamic pricing, AI can adjust prices in real time based on a multitude of factors, from competitor pricing to current demand levels and inventory availability. This capability ensures that prices are always optimized to maximize both sales and profit margins. AI doesn't just help in determining the right price but can also help forecast how pricing changes might impact demand, allowing for more strategic planning.

Of course, no discussion of AI in supply chain management would be complete without mentioning the integration of Internet of Things (IoT) devices. IoT sensors, embedded throughout the supply chain, can provide real-time data on everything from temperature and humidity to vehicle location and machine health. AI systems can process this data to provide actionable insights, from predicting equipment failures before they happen to optimizing environmental conditions for perishable goods.

AI in supply chain management is not without its challenges. Data privacy and security remain paramount concerns, especially with the vast amounts of data being processed. Companies must ensure that their AI systems comply with regulatory standards and that sensitive information is adequately protected. Integrating AI into existing supply chain systems can also be a complex task, requiring significant investments in technology and expertise.

Nevertheless, the benefits far outweigh the challenges. The capacity for real-time decision-making, enhanced efficiency, increased accuracy, and cost reductions transform the entire retail supply chain into a smoother, more agile operation. What's exciting is that we're only scratching the surface of what's possible. With advancements in AI technologies, the future holds even more promise for innovations that could reshape supply chain management in ways we can only begin to imagine.

On a broader scale, the implementation of AI in supply chain management also cultivates resilience. Natural disasters, political instabilities, and other unforeseen events can disrupt supply chains. AI systems can simulate various scenarios to prepare contingency plans, ensuring that companies are better equipped to handle disruptions and maintain continuous operations.

In conclusion, AI in supply chain management is more than just a technological upgrade; it's a paradigm shift that redefines efficiency, transparency, and resilience in the retail ecosystem. As companies continue to harness the power of AI, they'll find themselves better equipped to navigate the complexities of modern retail, delivering superior customer experiences while optimizing operational performance. The future of retail supply chains is bright, powered by the limitless possibilities that AI brings to the table.

Smart Warehouses

The innovation of smart warehouses stands as one of the most transformative applications of AI in the retail and shopping industry. Driven by an amalgamation of artificial intelligence, machine learning, and the Internet of Things (IoT), smart warehouses are redefining how inventory is managed, orders are fulfilled, and overall operational efficiency is achieved. Imagine a sprawling warehouse where robots scoot around autonomously, shelving items, picking products, and prepping them for shipment—all with minimal human intervention.

One of the key elements that make smart warehouses so efficient is the use of AI-powered robotics. These robots, often orchestrated by complex algorithms, are capable of performing tasks that would otherwise require significant human labor. Whether it's picking items off the shelves, packaging them, or even transporting them to shipping areas, these robots can handle it all. They are not just fast; they are incredibly accurate, reducing the margin for human error substantially. As a result, the supply chain becomes more reliable, and customer satisfaction increases.

Moreover, AI-driven predictive analytics play a crucial role in inventory management within these smart warehouses. By analyzing vast amounts of data—ranging from historical sales trends to current market conditions—AI can predict which products will be in demand and when. This makes it possible to maintain optimal inventory levels, ensuring that popular items are always in stock while minimizing overstock for slower-moving goods. The reduced need for excess inventory not only frees up valuable warehouse space but also cuts down on resources spent on maintaining unused stock.

Another remarkable feature of smart warehouses is their ability to utilize advanced sensor technology for real-time monitoring. With IoT sensors deployed throughout the warehouse, it's possible to track the movement of goods, monitor environmental conditions like temperature and humidity, and even detect potential equipment malfunctions before they become critical. The data collected by these sensors is continuously analyzed by AI systems, enabling warehouse managers to make informed decisions based on real-time information.

Imagine integrating these IoT sensors with machine learning algorithms to predict maintenance needs for warehouse machinery. AI can analyze patterns in the data to forecast when a machine is likely to fail, allowing for preemptive maintenance and avoiding costly downtime. This predictive maintenance capability ensures that operations in the warehouse continue smoothly and efficiently, contributing to a more streamlined supply chain.

Furthermore, the use of AI in smart warehouses extends to automated guided vehicles (AGVs) and drones. AGVs are employed to transport goods around the warehouse autonomously, while drones can be used for inventory audits and inspections, reaching areas that are difficult for humans to access. These technologies not only speed up operations but also enhance safety by reducing the need for human workers to perform potentially hazardous tasks.

Machine learning algorithms also facilitate dynamic slotting, a process that determines the most efficient placement of products within the warehouse. By analyzing factors such as the frequency of item retrieval and seasonal demand variations, AI systems can assign optimal storage locations for different products. This significantly reduces the time it takes for robots and human workers to locate and retrieve items, speeding up the entire order fulfillment process.

Order processing and fulfillment are further enhanced through AI-driven systems that can automatically verify and validate orders as they come in. These systems cross-check orders against real-time inventory data, ensuring that the items being ordered are indeed in stock and can be shipped without delay. If discrepancies are found, the system can alert human supervisors for immediate action, drastically reducing the number of errors and returns.

Transitioning to smart warehouses also brings about significant energy efficiencies. AI systems can optimize the usage of lighting, heating, and cooling in different zones of the warehouse, depending on the time of day and the areas being actively used. For instance, sections not in immediate use can be dimmed or have their temperature adjusted to save on energy costs. These energy savings not only reduce operational costs but also contribute to a company's sustainability goals.

Employee welfare is another area where smart warehouses have a notable impact. With robots taking over repetitive and physically strenuous tasks, human workers can be redirected towards more strategic, value-adding activities. This not only enhances job satisfaction but also reduces the risk of workplace injuries. AI can even assist in training human workers by providing real-time feedback and guidance, making it easier for employees to adapt to their new roles and responsibilities within the smart warehouse environment.

Let's not overlook the role of data analytics in enhancing the performance of smart warehouses. AI systems collect and analyze a plethora of data points, producing actionable insights that can help in optimizing every aspect of warehouse operations. For example, data can reveal bottlenecks in the order fulfillment process, allowing managers to make targeted improvements. Additionally, by analyzing customer feedback and return data, warehouses can identify recurring issues with specific products, enabling retailers to address quality problems proactively.

The impact of smart warehouses extends beyond the confines of the warehouse itself and permeates the entire retail ecosystem. Faster order processing and fulfillment mean shorter delivery times, which is a significant advantage in the competitive e-commerce landscape where consumers increasingly expect rapid delivery. Retailers can leverage these efficiencies to offer same-day or even next-hour delivery services, providing a competitive edge.

A fascinating aspect of smart warehouses is their adaptability. Whether a retailer deals in electronics, apparel, or perishable goods, AI systems can be customized to meet the specific needs of different types of products. For instance, temperature-sensitive items can be stored and monitored in climate-controlled sections, with AI ensuring that optimal conditions are consistently maintained. This adaptability makes smart warehouses a versatile solution for a variety of retail sectors.

In essence, smart warehouses embody the future of retail logistics. They represent a paradigm shift from traditional, labor-intensive warehousing to a model that is agile, efficient, and scalable. The benefits of adopting AI-driven smart warehouses are manifold, ranging from increased operational efficiencies and cost savings to enhanced customer satisfaction and reduced environmental impact. As AI technology continues to evolve, we can expect even more innovative solutions to emerge, further revolutionizing the landscape of retail and shopping.

CHAPTER 10:
AI IN WORK AND PRODUCTIVITY

In today's fast-paced work environment, AI is transforming the way we handle tasks and boost productivity. Imagine an intelligent task management system that predicts your priorities and allocates your time effectively, or a project management tool that processes massive datasets to keep projects on track. Virtual assistants are now more sophisticated, seamlessly integrating with your workflow to manage emails, schedule meetings, and even generate reports. These innovations reduce redundancy, enhance decision-making, and free up time for more creative endeavors. As AI continues to evolve, the potential for optimizing workplace efficiency grows exponentially, making our professional lives not only more productive but also more fulfilling.

Intelligent Task Management

In the modern workplace, where the pace of business can be dizzying, the ability to effectively manage tasks is a game-changer. Artificial Intelligence (AI) promises not just to streamline task management but to revolutionize it. Think of intelligent task management as your ultimate productivity booster, using AI to anticipate needs, allocate resources, and prioritize tasks in ways that humans simply can't match.

Imagine starting your workday with an automated briefing. Your AI assistant has already analyzed your emails, meetings, and project deadlines. It then generates a prioritized list of tasks tailored to your work habits and objectives. You're not just managing tasks; you're orchestrating an intelligently curated schedule designed for peak efficiency.

This isn't a futuristic fantasy; it's happening now. Tools driven by natural language processing (NLP) and machine learning (ML) are already in use. They analyze your to-do lists, email patterns, and even your social interactions to optimize your workflow. For instance, if you have a habit of checking emails first thing in the morning, the AI might suggest critical emails to address while holding off flagging less urgent ones until later.

The role of AI in intelligent task management goes beyond mere scheduling. It taps into predictive analytics to foresee potential roadblocks and bottlenecks. For example, if you're working on a project that requires input from multiple team members, AI can predict delays based on past performance and suggest ways to mitigate them.

But how exactly does AI anticipate your needs? Through data. Tons of it. By analyzing historical data, AI can identify patterns that are likely to recur. So, if certain tasks consistently take longer on Mondays, the AI will adjust deadlines and resource allocations accordingly. It learns from every interaction, becoming smarter and more intuitive the longer you use it.

One pivotal aspect of intelligent task management is its ability to integrate with various platforms. Whether you use Google Workspace, Microsoft Office, or other project management tools, AI-driven systems can seamlessly pull information from these diverse sources. This creates a unified dashboard from which you can manage all aspects of your work, reducing the need to juggle multiple apps and platforms.

Moreover, AI in task management can foster collaboration. In team settings, it can allocate tasks based on each member's strengths and workloads, ensuring that no one is overwhelmed or underutilized.

By continuously monitoring progress and adjusting priorities, AI keeps everyone on track, hitting milestones like clockwork.

Consider the mental load this relieves. With AI handling the nitty-gritty details, you're freed to focus on high-level thinking and creative problem-solving. This shift isn't merely about efficiency; it's about unlocking cognitive potential. You'll likely find that you're not only getting more done but also producing higher quality work.

Another compelling feature is AI's capability for real-time adjustments. Suppose a new, high-priority task drops into your lap mid-day. AI can reconfigure your task list, slotting the urgent task in and bumping less critical ones, all while notifying you of the changes. This real-time dynamism ensures that you're always working on what matters most without the stress of constant manual re-prioritization.

In essence, intelligent task management brings a level of personalization that static systems simply can't offer. Your unique working style and preferences are continually fed back into the system, fine-tuning it to better serve you. Imagine an assistant who not only organizes your work but also learns from you to become more effective over time.

The financial benefits are also significant. Businesses that implement AI-driven task management systems report marked increases in productivity and employee satisfaction, translating into improved bottom lines. Time saved on manual task scheduling and prioritization is time that can be spent on value-generating activities.

Think about meetings for a moment. AI can schedule them at optimal times based on participants' availability and productivity patterns. It can also send out agendas, prompt attendees for preparation, and even summarize action points afterward. Thus, meetings become another arena where smart task management reigns supreme.

As more enterprises begin adopting this cutting-edge technology, we're likely to see an evolving landscape where AI becomes an indispensable co-worker. The transition to intelligent task management won't be without its challenges, but the long-term gains in efficiency, collaboration, and employee satisfaction will far outweigh the initial implementation efforts.

In summary, intelligent task management is more than an incremental improvement; it's a paradigm shift. It leverages vast amounts of data to offer insights and efficiencies that were previously unimaginable. By learning from historical patterns and integrating seamlessly with existing tools, AI in task management stands to transform how we approach work fundamentally. The result is not just a more productive you but a smarter, happier, and more fulfilled worker, ready to tackle the complexities of the modern workspace.

AI in Project Management

Harnessing the power of artificial intelligence in project management can be transformative, adding efficiency, accuracy, and insight into a traditionally human-driven discipline. Project management has always required meticulous planning, careful execution, and the ability to pivot as situations change. Integrating AI into this mix enhances it by ensuring tasks and resources are optimally managed, risks are effectively mitigated, and project outcomes are more predictable.

One of the most significant contributions of AI in project management is intelligent task management. Traditional project management tools often require manual input and oversight, but AI can automate these processes to a large extent. For instance, AI algorithms can predict the timeframe and resources required for different tasks, automatically assign team members based on their skill sets and availability, and adjust timelines dynamically as variables change. This automation not only reduces the workload of project managers but also minimizes human error.

Imagine a project where timelines are continually updated in real time, and tasks are redistributed based on the current workload and performance of team members. AI can analyze historical data to anticipate bottlenecks and suggest preventive measures, ensuring that the project runs smoothly. This predictive capability can be a game-changer, especially for complex projects involving multiple stakeholders and tight deadlines.

Communication within a project team is crucial, and AI significantly improves this aspect by providing intelligent communication tools. Natural language processing (NLP) allows AI to understand and generate human language, making collaboration tools more intuitive and effective. AI-driven chatbots can handle routine queries, schedule meetings, and even draft emails, freeing up time for more strategic activities.

AI also enhances decision-making processes. By aggregating and analyzing vast amounts of data, AI can provide insights that would be impossible to glean manually. For instance, an AI system can identify patterns and correlations that might not be immediately obvious to human analysts. These insights can inform decisions regarding budget allocation, risk management, and resource distribution, ensuring that projects are grounded in data-driven strategies.

Risk management is another area where AI has a profound impact. Predictive analytics can identify potential risks early in the project lifecycle, allowing teams to take proactive measures. AI can assess risks based on historical data, current project parameters, and external factors such as market trends or economic conditions. This proactive approach to risk management helps in mitigating threats before they turn into crises.

Moreover, AI facilitates continuous learning and improvement within project management. Machine learning algorithms can analyze outcomes of completed projects to understand what worked well and what didn't. This continuous feedback loop allows organizations to refine their project management practices, leading to more consistent success over time. Teams can learn from previous projects and avoid past mistakes, creating a culture of continuous improvement.

Resource management also benefits significantly from AI integration. AI can optimize the allocation of resources, ensuring that human talent, time, and materials are used most efficiently. For example, AI can match team members with the tasks that best fit their skills and experience, thus maximizing productivity and job satisfaction. Additionally, AI can monitor the utilization of physical resources, reducing waste and ensuring that everything is used optimally.

The integration of virtual assistants in project management tools adds another layer of efficiency. These AI-driven assistants can perform a variety of tasks, from setting up project timelines to tracking progress and generating reports. They can interact with team members through voice or text, providing assistance whenever needed. By handling routine administrative tasks, virtual assistants allow project managers and team members to focus on more strategic elements of the project.

In addition to managing tasks and resources, AI can also enhance strategic planning. Algorithms can analyze market trends, competitor activities, and internal capabilities to provide comprehensive strategies for achieving project goals. This level of strategic insight can be particularly valuable in competitive markets where agility and informed decision-making are essential.

AI algorithms can also encourage better adherence to compliance and regulatory requirements by monitoring project activities and flagging potential issues. This ensures that projects are not only completed on time and within budget but also in line with industry standards and regulations. AI can keep track of documentation, deadlines, and compliance milestones, reducing the risk of oversights that could have legal or financial repercussions.

Ultimately, the use of AI in project management is not about replacing human project managers; it is about augmenting their capabilities. AI provides tools and insights that empower project managers to make better decisions, work more efficiently, and deliver better results. By automating routine tasks and providing data-driven insights, AI allows project managers to focus on the aspects of their work that require human creativity, judgment, and interpersonal skills.

In conclusion, AI is revolutionizing project management by introducing intelligent task management, enhancing communication, improving decision-making, facilitating risk management, and optimizing resource allocation. The ability to harness AI's power to analyze data, predict outcomes, and automate routine tasks can transform how projects are managed, leading to greater efficiency, reduced costs, and more successful outcomes. As AI continues to advance, its role in project management will undoubtedly grow, making it an indispensable tool for organizations striving for excellence in their projects.

Virtual Assistants

Virtual assistants are rapidly transforming how we approach daily tasks and professional responsibilities. Fueled by the growth of artificial intelligence, these tools blend seamlessly into our workflows, providing unparalleled support in both personal and work-related activities. They leverage algorithms to learn from our habits, preferences, and routines, offering solutions that save time, enhance efficiency, and boost productivity.

Virtual assistants began as basic tools for setting reminders or answering queries but have evolved into sophisticated systems capable of managing complex tasks. These assistants can schedule meetings, manage emails, conduct research, and even handle customer service inquiries. The escalation in their capabilities is largely due to advancements in natural language processing (NLP) and machine learning, making them more intuitive and conversational.

In the realm of work and productivity, virtual assistants have become indispensable. For example, they can prioritize and organize tasks based on deadlines and levels of importance. By integrating with calendar applications and project management software, they ensure that nothing falls through the cracks. This allows professionals to focus on more strategic aspects of their roles, fostering a more productive work environment.

One of the notable advantages of virtual assistants is their ability to act as knowledge workers. Need to draft a report or analyze data quickly? A virtual assistant can do that. Through machine learning, these assistants can process and interpret vast amounts of data, providing actionable insights much faster than a human could.

Another key benefit is in communication management. Virtual assistants can sift through your emails, prioritize messages, and even craft responses. This is particularly useful in high-paced work environments where timely communication is crucial. They can flag important emails, alerting you to matters that need immediate attention, and handle routine inquiries autonomously.

Virtual assistants also offer a significant edge in managing virtual meetings. They can schedule meetings across different time zones, book conference rooms, and set up video calls. Some can even take notes during meetings, ensuring that all key points and action items are captured. This reduces the administrative burden on employees, allowing them to concentrate on the core agenda.

In project management, virtual assistants are proving to be game-changers. They can track project milestones, send reminders for upcoming deadlines, and keep team members updated on their tasks.

By offering real-time project tracking, these assistants ensure that everyone is on the same page, thereby minimizing delays and enhancing collaboration.

Moreover, virtual assistants can be programmed to align with company goals and workflows. They can integrate into existing systems like CRM software, making it easier for sales and customer service teams to track interactions and follow up with clients. Automating these processes not only saves time but also improves accuracy and accountability.

In addition to supporting everyday tasks, virtual assistants are becoming more adept at context-aware computing. This means they can understand the context of your requests, making their assistance more relevant and timely. For instance, if you're preparing for a meeting with a particular client, your virtual assistant can pull up recent communications, relevant documents, and even suggest talking points, all in one go.

Another compelling aspect is the customization options available with virtual assistants. You can tailor them to meet specific needs, whether that's integrating them with specialized software or configuring them to handle unique tasks. This level of personalization ensures that the assistant aligns perfectly with your workflow, maximizing efficiency.

Security is another critical consideration, and modern virtual assistants incorporate robust measures to protect sensitive information. They use encryption, multi-factor authentication, and other advanced technologies to ensure that data remains confidential. This is particularly important in workplaces that handle sensitive or regulated information.

Additionally, the user interface of virtual assistants has seen significant improvements. Most now feature intuitive dashboards that offer a visual representation of tasks, deadlines, and priorities. This makes it easier to manage tasks and track progress, providing a clear overview of your workload at a glance.

Looking ahead, the future of virtual assistants in enhancing work and productivity seems incredibly promising. As AI technologies continue to advance, we can expect these tools to become even more integrated into our daily lives, offering deeper insights and more intelligent automation capabilities.

In conclusion, virtual assistants are not just a novelty; they've become essential tools that offer tangible benefits in enhancing productivity and efficiency at work. By leveraging the power of AI, they free up time and cognitive resources, allowing professionals to focus on what truly matters. As these technologies continue to evolve, the potential for further enhancements in productivity seems limitless.

CHAPTER 11:
AI IN SOCIAL MEDIA

Social media platforms have become an integral part of our everyday lives, and AI is revolutionizing how we interact within these digital landscapes. From tailoring content to match user preferences to leveraging sophisticated analytics for deeper insights, AI ensures that every scroll and click is more relevant and engaging. Automated bots assist in customer service, respond to queries, and manage online communities, freeing up human time for more complex tasks. This interplay of AI and social media fosters a more personalized, efficient, and responsive online experience, creating communities that are not only more connected but also more intelligent.

Content Personalization

Content personalization in social media isn't just a buzzword; it's a technological transformation that's redefining how we consume information. Think about the last time you scrolled through your social media feed. Did you notice how the posts, advertisements, and recommendations seemed almost tailor-made for you? This isn't a coincidence. It's the result of intricate algorithms powered by Artificial Intelligence (AI) working behind the scenes to learn your preferences, behaviors, and interests.

At its core, content personalization leverages vast amounts of data to provide a customized experience to each user. AI systems analyze various factors: your past interactions, likes, shares, comments, time spent on specific posts, and even the type of device you are using. By doing so, they can predict what content you will find engaging and display those items more prominently in your feed.

One of the primary techniques used in content personalization is collaborative filtering. This method involves analyzing user behaviors and finding patterns among different users. For example, if you like and frequently engage with posts about cooking, the AI will suggest content liked by others who have similar interests. This helps create a web of connectivity and relevance that keeps you engaged and immersed in the platform.

Another critical technique is content-based filtering, which focuses on the attributes of the content itself. Here, AI analyzes the type of content that you have interacted with and seeks out similar posts. For example, if you often watch cooking videos, the AI will scan metadata, keywords, and tags associated with those videos to offer you more content of a similar nature. This ensures that your feed is filled with posts that cater to your unique tastes and preferences.

Furthermore, AI systems continually evolve based on real-time data. This dynamic adaptability ensures that the content remains relevant as your interests change over time. For instance, if you start liking and engaging with fitness-related posts more frequently, AI will pick up on this shift and adjust your feed to show more fitness content. The continuous feedback loop allows for an ever-evolving, highly personalized user experience.

It's not just about content; ads are heavily personalized too. Businesses use AI-driven analytics to target potential customers with pinpoint accuracy. By analyzing your behavior on social media and even outside it, AI can serve up ads that align closely with your preferences and needs. This means less irrelevant advertising cluttering your feed and more products and services that might genuinely interest you.

But the influence of AI in content personalization doesn't stop at what you see on your feed. It goes deeper, affecting how content creators produce and share their work. AI tools provide influencers and content creators with insights into what works best for their audience. By analyzing engagement metrics, these tools help creators tailor their content to meet the expectations and preferences of their followers, ensuring higher engagement and satisfaction.

Moreover, these personalized experiences extend into the realm of news and information. Instead of being flooded with a generic feed of articles, AI curates a selection that matches your reading habits and interests. This helps you stay informed on the topics that matter most to you without having to sift through irrelevant content. However, this also brings into focus the issue of echo chambers, where users are exposed predominantly to information that reinforces their existing beliefs, potentially limiting broader perspectives.

The power of AI in content personalization can also be seen in the way social media platforms facilitate human connections. Recommendations for new friends, groups, and pages are all driven by sophisticated algorithms designed to enhance your social media experience. By suggesting connections that align with your interests and behaviors, these platforms help foster meaningful interactions and build communities.

Social media giants like Facebook, Instagram, and Twitter are continually refining their algorithms to enhance the personalization experience. Take Spotify, for example, which uses a combination of collaborative filtering, Natural Language Processing (NLP), and audio analysis to recommend music that you didn't even know you wanted to hear. Similarly, Netflix employs advanced machine learning techniques to suggest movies and shows based on your viewing history. These AI-driven enhancements ensure that users spend more time on these platforms because they're continuously engaged with content that resonates.

Moreover, AI's role in content personalization is also paving the way for more accessible and inclusive social media experiences. By analyzing user interactions and feedback, AI can suggest improvements and modifications that make platforms more user-friendly for individuals with disabilities. This includes features like automatic captions, voice commands, and tailored content delivery that breaks down barriers and creates more inclusive digital spaces.

But like all powerful tools, AI-driven content personalization comes with its own set of challenges and ethical considerations. The vast amount of data required for personalization raises significant privacy concerns. Users must trust that their data is being collected and used responsibly. Furthermore, there is the risk of perpetuating biases that can exacerbate inequalities. Without careful oversight, AI systems could inadvertently reinforce harmful stereotypes or marginalize certain groups based on the data they are trained on.

Despite these challenges, the potential benefits of AI-driven content personalization in social media are vast. By offering a more engaging and relevant user experience, these technologies can enhance our digital interactions and make social media a more enjoyable space. For brands and businesses, the ability to deliver personalized content can lead to more effective marketing strategies, higher engagement rates, and ultimately, a better ROI.

Looking to the future, AI in content personalization will likely become even more sophisticated. Advances in machine learning, deep learning, and data analytics will allow for increasingly nuanced and accurate personalizations. Imagine a social media feed that understands your mood and delivers content to uplift or inform you at the right moments. Or a platform that predicts your needs before you even realize them, suggesting solutions and content that enhance your daily life.

William Scott

In conclusion, content personalization through AI is a powerful tool capable of transforming how we interact with social media. While it brings about remarkable enhancements in user experience, it also necessitates a balanced approach to ethical considerations and data privacy. As this technology continues to evolve, it promises a future where digital interactions are not only more engaging but also more meaningful and enriching.

AI-Driven Analytics

In the dynamic world of social media, data is king. Every post, comment, like, and share generates a wealth of information. But, interpreting this data manually would be an onerous task. Enter AI-driven analytics—a game-changer in deciphering the vast labyrinth of social media interactions.

AI-driven analytics leverage machine learning algorithms to process and analyze volumes of data that would be insurmountable for humans alone. These algorithms can quickly identify patterns, trends, and anomalies, providing actionable insights that were previously lost in the noise. Imagine being able to understand user behavior, sentiment, and engagement in real-time. That's the power of AI analytics.

Social media platforms like Facebook, Twitter, and Instagram use AI-driven analytics to enhance user experience and optimize content distribution. By analyzing user interactions, AI can predict what types of content will likely engage specific users and tailor their feeds accordingly. This personalized approach not only boosts user satisfaction but also maximizes engagement metrics, creating a win-win scenario.

Businesses significantly benefit from AI-driven analytics in social media. For example, sentiment analysis tools can evaluate public opinion about a brand or product by analyzing social media mentions. Whether it's a new product launch or a PR crisis, these tools offer real-time insights that guide decision-makers in crafting appropriate re-

120

sponses and strategies. Moreover, insights gained from AI analytics can inform product development, marketing campaigns, and customer service improvements.

Another remarkable application lies in trend forecasting. AI algorithms can sift through enormous amounts of data to identify emerging trends before they become mainstream. By recognizing these patterns early, businesses can stay ahead of the curve, ensuring they're always relevant and up-to-date. This is particularly crucial in industries like fashion, technology, and entertainment, where trends evolve rapidly.

AI-driven analytics also play a crucial role in monitoring and maintaining brand reputation. Automated systems can flag negative comments or reviews instantly, allowing businesses to address issues promptly and publicly showcase their customer service efforts. On a larger scale, these systems can monitor the overall sentiment toward a brand over time, providing valuable feedback on long-term strategies.

Aside from corporate uses, AI-driven analytics enhance individual user experiences. Personalized recommendations are a prime example. Whether suggesting new friends, groups to join, or pages to follow, AI algorithms work tirelessly behind the scenes, ensuring content aligns with individual preferences. This creates an engaging, tailored social media environment that keeps users coming back.

Political campaigns leverage AI-driven analytics in unique ways as well. By analyzing public sentiment and engagement metrics, campaigns can fine-tune their messaging to resonate more deeply with voters. This data-driven approach enables more targeted and effective communication strategies, which are crucial in the fast-paced, highly polarized world of modern politics.

One fascinating, albeit controversial, use of AI in social media is its application in surveillance and law enforcement. By monitoring social media for specific keywords or behavioral patterns, AI algorithms can flag potential threats or criminal activities. While this raises significant privacy and ethical concerns, it demonstrates the far-reaching impact of AI-driven analytics in even the most unexpected areas.

Influencer marketing has seen a transformation thanks to AI analytics as well. Identifying the right influencers for a brand used to be a cumbersome process, often based more on intuition than data. Now, AI can analyze metrics like engagement rates, audience demographics, and past performance to find the perfect match. This precision helps brands collaborate with influencers who genuinely align with their identity and goals, making campaigns more effective.

AI-driven analytics also contribute to tackling misinformation and fake news on social media. Algorithms can assess the credibility of sources, cross-reference facts, and even predict the likelihood of content being false or misleading. By flagging suspicious posts for human review, these systems help maintain the platform's integrity and reliability.

The realm of customer service has been revolutionized as well. With AI-driven analytics, companies can monitor customer interactions across social media platforms, analyzing complaints, queries, and feedback. This data is invaluable for improving customer service protocols and training, as it provides a direct line to what the consumers are saying and feeling.

Even nonprofit organizations and social causes benefit immensely from AI-driven analytics. By understanding what motivates their audience, which issues resonate most deeply, and which tactics drive the most engagement, these organizations can craft more compelling narratives and create impactful campaigns. This is crucial for maximizing reach and mobilizing resources effectively.

Monitoring the efficacy of social media campaigns is another area where AI-driven analytics shine. Traditional metrics like likes, shares, and comments only scratch the surface. AI can assess deeper engagement levels, such as the sentiment of comments or the virality potential of a post. This allows for a more nuanced and holistic understanding of a campaign's performance.

AI-driven analytics' impact doesn't end with external communication. Internally, companies can analyze employee social media usage to gauge morale and uncover issues before they escalate. While it's important to navigate this area with sensitivity and respect for privacy, the insights gained can be instrumental in shaping company culture and addressing employee concerns.

It's clear that AI-driven analytics offer something of immense value across various sectors. They provide a method to transform raw data into actionable, insightful information. This ability can inform decisions, fuel growth, and create more meaningful connections between users, businesses, and other entities.

All of this culminates in one undeniable truth: AI-driven analytics are not just a tool but a catalyst for change in the social media landscape. They enable us to navigate the complexities of human behavior and digital interaction with unprecedented clarity. Whether it's personalizing user experiences, optimizing business strategies, or ensuring safety and fairness, AI-driven analytics stand at the forefront, making social media not just a platform for interaction, but a domain of innovation and improvement.

Bots and Automation

Bots and automation are integral components of AI in the realm of social media. Their omnipresence is felt across various platforms, creating ripple effects that fundamentally alter how we interact with digital spaces. At their core, bots are software applications that run automated tasks (scripts) over the internet. They perform simple and repetitive tasks far faster than human users could ever manage. On social media, bots have assumed tasks that range from posting regular updates to engaging with users and even monitoring social media traffic.

One of the most obvious applications is customer service. Businesses deploy bots to respond to customer queries in real-time, providing instant information and solutions to consumers without the need for human intervention. For example, a customer might message a company's Facebook page with a question about store hours. The bot can instantly respond with accurate information, ensuring that the consumer's needs are met promptly. This not only enhances user experience but also significantly reduces the workload on human customer service representatives.

Automation in social media doesn't stop at customer service. It's also employed in managing social media accounts more efficiently. Social media managers use bots to schedule posts, follow users back, and share relevant content across various platforms. By automating these tasks, they can devote more time to creative aspects of social media strategy, such as crafting unique content or engaging meaningfully with followers.

Content creation is another area where AI-driven bots shine. Bots can generate news stories, write articles, or even create social media posts. Leveraging natural language processing (NLP), these bots can analyze trending topics and craft posts that align with current internet buzz. This is particularly useful for news outlets and blogs that need to keep their content fresh and relevant.

AI in social media also involves the use of chatbots, which can hold conversation-like interactions with users. These chatbots are equipped with NLP capabilities, enabling them to understand and respond to user inputs in a way that mimics human conversation. Businesses often utilize chatbots for marketing purposes, creating interactive campaigns that capture user interest through quizzes, polls, and personalized recommendations.

An often-overlooked benefit of bots and automation is sentiment analysis. Bots can sift through vast amounts of social media data to gauge public sentiment about a brand, product, or individual. By analyzing keywords and the context of conversations, bots can provide insights into how a brand is perceived and signal potential PR crises before they escalate.

Despite their benefits, bots and automation come with their own set of challenges. One significant issue is the potential for misuse. Malicious actors can deploy bots for nefarious purposes such as spreading misinformation, generating fake reviews, or artificially inflating follower counts. These activities not only undermine the integrity of social media platforms but also pose ethical dilemmas that need addressing.

Platform developers are continually updating algorithms to identify and weed out malicious bots. For instance, Twitter regularly purges accounts that exhibit bot-like behavior to maintain the platform's integrity. On a broader scale, the use of machine learning models helps in distinguishing harmful bots from benign ones by analyzing patterns in behavior and interaction.

Automation also touches on ethical concerns around job displacement. As more tasks become automated, there's a growing worry about the future of jobs traditionally held by humans. Social media strategists and community managers might find parts of their roles increasingly overtaken by bots. This underscores the importance of upskilling and evolving human roles to focus on tasks that require creative and strategic thinking – areas where AI still lags behind human capability.

In conclusion, bots and automation in social media present a double-edged sword. On one side, they offer unprecedented efficiencies and capabilities that significantly enhance user interaction, customer service, and content management. On the other, they introduce ethical and operational challenges that necessitate careful consideration and management. As AI continues to evolve, the balance between leveraging its advantages and managing its pitfalls will be crucial in shaping the future of social media landscapes.

CHAPTER 12:
AI IN REAL ESTATE

The integration of AI in real estate is reshaping the industry, making processes more efficient and user-friendly for buyers, sellers, and managers alike. Imagine a world where AI-powered algorithms analyze market trends to provide accurate property valuations or match buyers with homes that fit their preferences like never before. Smart property management systems using machine learning continuously monitor building conditions and automatically schedule maintenance, reducing costs and improving tenant satisfaction. Virtual tours, enhanced by AI, allow potential buyers to explore properties remotely, experiencing realistic walkthroughs that save time and streamline decision-making. Through these advanced technologies, the real estate sector is rapidly evolving, offering a smoother, more intuitive experience for everyone involved.

Smart Property Management

As urbanization continues to rise, property management faces increasing complexities, from routine maintenance to tenant satisfaction. Enter Artificial Intelligence (AI), poised to revolutionize how properties are managed, enhancing efficiency and providing comprehensive solutions to previously cumbersome tasks. Smart property management leverages AI technologies to streamline operations, cut costs, and improve overall tenant experiences.

Imagine a building where the lights adjust automatically based on occupancy and natural light levels, or a heating system learns and mimics tenant preferences to optimize energy consumption. AI-driven systems can also predict when maintenance issues are likely to occur, allowing property managers to address concerns before they escalate into costly repairs. This doesn't just save money; it also ensures that properties are kept in optimal condition, thereby retaining tenant satisfaction and potentially increasing property value.

The implementation of AI in property management begins with smart sensors and Internet of Things (IoT) devices. These connected devices collect vast amounts of data, from temperature and humidity levels to motion detection and appliance efficiency. AI algorithms analyze this data, making it possible to automate various tasks. For example, smart HVAC systems can adjust temperature settings in real-time based on room occupancy, improving energy efficiency and comfort.

But AI's role in smart property management goes beyond environmental control. It's transforming security as well. Modern security systems equipped with AI can identify and differentiate between residents, visitors, and potential intruders using facial recognition technology. Combined with predictive analytics, these systems can even anticipate security breaches before they happen, providing a proactive approach to safety.

AI-powered platforms significantly enhance communication between property managers and tenants. Chatbots, for instance, can handle routine inquiries about rent payments, lease details, or maintenance requests around the clock, freeing up property managers to focus on more complex tasks. Natural Language Processing (NLP) capabilities enable these chatbots to understand and respond to tenant queries in a conversation style that feels almost human.

Beyond tenant interactions, AI enhances the efficiency of administrative tasks. Automated scheduling systems can manage property tours, routine inspections, and lease renewals, ensuring that nothing slips through the cracks. AI can also assist in financial management by monitoring income and expenditure, identifying discrepancies, and even predicting future financial performance based on historical data.

AI's ability to analyze vast datasets allows it to identify trends and forecast market behavior with a level of accuracy that was previously unachievable. Property managers can use AI to determine optimal times for rental adjustments, identify underperforming assets, and even guide investment strategies.

The utilization of AI in facilities management also grows as operational requirements become more sophisticated. AI-driven platforms can predict equipment failures by continuously monitoring machinery performance and flagging anomalies before they result in breakdowns. This predictive maintenance minimizes downtime and extends the lifespan of critical infrastructure.

Environmental sustainability is another area where AI-driven property management shines. By optimizing energy use, water consumption, and waste management, smart buildings can significantly reduce their environmental footprint. AI algorithms can identify patterns in resource usage and suggest efficiency improvements that align with sustainability goals.

Tenant satisfaction remains a top priority, and AI offers innovative ways to enhance it. AI can tailor tenant experiences by learning preferences and automating personalized settings, whether adjusting living environments or curating localized services and amenities. Such personalized interactions lead to higher tenant retention rates, a critical factor in maintaining stable revenue streams.

The ease of integrating AI innovations into existing property management systems is a critical advantage. Many AI platforms are designed to be modular and scalable, allowing gradual implementation without significant upfront costs or disruptions. Property managers can start by integrating AI into one aspect, such as energy management, and then expand to other areas as benefits become evident.

Challenges do exist, particularly regarding data privacy and security. Keeping sensitive tenant information secure is paramount, and AI-powered systems must adhere to stringent data protection standards. However, advancements in AI-driven cybersecurity are continually evolving, providing robust defenses against potential breaches.

Looking ahead, the role of AI in property management is set to expand even further. The convergence of AI with other emerging technologies, like blockchain and augmented reality (AR), offers new possibilities. Consider blockchain's potential for ensuring transparent and fraud-resistant transactions or AR for immersive property tours. Such integrations will make property management more innovative and more efficient, secure, and engaging.

In conclusion, AI has ushered in a new era of smart property management that prioritizes efficiency, security, sustainability, and tenant satisfaction. By harnessing the power of AI, property managers can navigate the complex landscape of modern real estate with unprecedented precision and insight. The potential benefits are substantial, making AI an indispensable tool in the evolving field of property management.

AI in Home Buying and Selling

The real estate market has historically been a labyrinth of complex processes and paperwork. From searching for the perfect home to closing the deal, the journey is fraught with hurdles. However, Artificial Intelligence (AI) is reshaping this landscape, making home buying and selling more straightforward and efficient. Gone are the days when buyers and sellers had to rely solely on real estate agents for every bit of advice. Today, AI is providing tools that are democratizing access to real estate insights, enhancing customer experiences, and streamlining transactions.

One of the critical ways AI is transforming home buying is through predictive analytics. By analyzing vast amounts of data, AI algorithms can predict property values with remarkable accuracy. These algorithms consider numerous factors, including historical selling prices, market trends, neighborhood amenities, and even the quality of local schools. As a result, buyers can make more informed decisions, and sellers can set realistic prices.

Moreover, AI-powered platforms like Zillow and Redfin leverage machine learning to offer personalized home recommendations. These platforms analyze a user's search history, preferences, and behaviors to suggest properties that match their criteria. This personalized experience saves time and ensures that buyers are more likely to find a home that meets their needs. It's akin to having a dedicated real estate agent who understands your preferences better than anyone.

But AI doesn't just assist in searching for homes; it also aids in the negotiation process. AI tools can analyze market conditions, comparable property sales, and even sentiment analysis from social media and news articles to provide real-time insights on the best time to buy or sell. This empowers buyers and sellers with data-driven advice, potentially leading to better deals and more favorable outcomes.

On the selling side, AI technology is streamlining the listing process. Traditional methods of listing homes on the market can be time-consuming and often miss key selling points. AI-driven tools can automatically generate compelling property descriptions, highlight unique features, and even suggest optimal listing prices based on market analysis. This not only speeds up the process but also ensures that listings are more attractive to potential buyers.

When it comes to the actual transaction, AI-driven platforms are simplifying the paperwork and legalities involved in buying and selling a home. Smart contracts, powered by blockchain and AI, are making transactions more secure and transparent. These digital contracts automate the verification of terms and the execution of agreements, reducing the risk of fraud and minimizing the need for intermediaries. Buyers and sellers can have more confidence that their transactions will proceed smoothly and securely.

Virtual tours have become increasingly popular, especially in the wake of global events that limit physical interactions. AI technologies like computer vision and augmented reality enable potential buyers to take immersive virtual tours of properties. These tours can be conducted from the comfort of one's home, offering a detailed view of the property's layout, design, and features. This not only saves time but also allows for a more flexible and convenient home-buying process. Imagine walking through a house on the other side of the country without leaving your living room.

Another remarkable development is the use of AI in property management for sellers who are looking to enhance the appeal of their homes. AI tools can provide suggestions for renovations and improvements that could increase the property's value. By analyzing market trends and buyer preferences, these tools can recommend changes that are most likely to resonate with potential buyers. This might include anything from updating a kitchen to adding smart home features, making the property more attractive and market-ready.

Then there's the aspect of understanding buyer behavior. By utilizing natural language processing and sentiment analysis, AI systems can gauge buyer reactions to different homes and listings. These insights help real estate agents tailor their strategies, whether it's tweaking a marketing campaign or adjusting the asking price. In essence, AI allows for a more responsive and adaptive approach to selling homes.

AI is also playing a crucial role in enhancing customer service within the real estate industry. Virtual assistants, such as chatbots, are available 24/7 to answer questions, schedule viewings, and provide detailed information about properties. These AI-driven agents can handle multiple inquiries simultaneously, ensuring that potential buyers receive prompt and accurate information. This level of service can significantly improve the buyer experience, making the home-buying process more pleasant and less stressful.

For real estate investors, AI provides tools for analyzing potential investments. Predictive analytics can assess the long-term value of properties, taking into account factors like urban development plans, economic conditions, and demographic shifts. This allows investors to make more informed decisions, maximizing their return on investment and minimizing risks.

In conclusion, AI is revolutionizing home buying and selling in myriad ways. From predictive analytics and personalized recommendations to virtual tours and smart contracts, the technology is streamlining every aspect of the process. By making data-driven insights accessible to everyone, AI is democratizing the real estate market, enabling both buyers and sellers to navigate it with greater confidence and efficiency. As AI continues to evolve, its impact on the real estate industry is likely to grow, offering even more innovative solutions to simplify and enhance the home buying and selling experience.

AI's integration into real estate is more than just a trend; it's a transformative force that is fundamentally changing how we buy and sell homes. By harnessing the power of AI, we can look forward to a future where the complexities of real estate are mitigated by smart, data-driven solutions that make the process smoother, more transparent, and ultimately more enjoyable for everyone involved.

Virtual Tours

In recent years, the real estate industry has undergone a revolutionary transformation, largely owing to the advancements in Artificial Intelligence (AI). One of the most impactful innovations in this domain is the rise of virtual tours. Whether you're a prospective homebuyer, a real estate agent, or a property manager, virtual tours have added a new layer of convenience and efficiency to the property viewing process.

The concept of virtual tours isn't entirely new, but AI has significantly enhanced its effectiveness and appeal. Traditional methods of property touring required physical presence, which often meant scheduling complexities, travel time, and sometimes an inconvenient rush-through. However, AI-powered virtual tours eliminate these hurdles. By enabling prospective buyers to explore properties from the comfort of their own homes, this technology brings the viewing process to an entirely new level.

Powered by sophisticated algorithms, virtual tours often incorporate 3D imaging, augmented reality (AR), and even virtual reality (VR). These technologies work in tandem to provide a realistic representation of the property. For instance, 3D imaging enables detailed and precise models of the space, allowing viewers to grasp the layout and dimensions accurately. Augmented reality adds a layer of interactivity, where users can visualize how a room might look with different furniture or decor.

Imagine you're relocating to a new city and can't afford multiple trips to scout properties. With a few clicks, you can virtually walk through multiple homes, inspecting every nook and cranny. AI algorithms analyze your preferences and can even recommend properties that match your tastes, making your search more efficient.

But the utility of virtual tours doesn't stop with individual buyers. Real estate agents also benefit significantly. The ability to showcase properties digitally means they can market to a broader audience without the constraints of geography. Furthermore, AI can assist agents in curating personalized property portfolios for clients, improving the chances of a sale.

One of the most exciting aspects of AI-enhanced virtual tours is the integration of machine learning. Over time, the system learns from user interactions, refining and improving recommendations. This not only personalizes the viewing experience but also provides invaluable analytics to real estate professionals. For example, data on which rooms are frequently viewed or which features attract the most interest can inform future property listings and marketing strategies.

Moreover, virtual tours offer unparalleled accessibility. People with disabilities, for instance, find it easier to view properties virtually. Navigating a house with stairs or narrow hallways is no longer a barrier when the viewing is done online. This universality serves to make the real estate market more inclusive.

From a seller's perspective, virtual tours can significantly reduce the time a property spends on the market. Listings with virtual tours receive more views and generate higher interest as prospective buyers appreciate the ability to explore the home in detail before committing to a physical visit. This pre-screening process attracts more serious buyers to open houses and significantly increases the likelihood of a quicker sale.

In addition to visualizing the property, some advanced virtual tour platforms now offer AI-driven narrative guides. As you navigate through the digital space, an AI-generated voice provides contextual information about the home's features, neighborhood amenities, and even local school ratings. This level of depth can foster a more engaging and informed viewing experience, almost akin to having a personal tour guide walking alongside you.

However, the implementation of virtual tours isn't without its challenges. Privacy and data security are paramount concerns. Ensuring that users' data is protected and maintaining the privacy of the property owners is crucial. AI developers are continually working on securing these platforms to prevent unauthorized access and data breaches.

Another consideration is the digital divide. Not everyone has access to high-speed internet or the latest technology required to benefit from AI-powered virtual tours. Developers and real estate professionals must strive to make these tools accessible and user-friendly for everyone.

Despite these challenges, the potential of virtual tours in the real estate sector is immense. They're not just a temporary solution born out of necessity; they're shaping up to be a staple in the future of property marketing and sales. Imagine a future where you can attend a live virtual open house with a real estate agent available to answer questions in real-time, or perhaps where AI could predict your housing needs based on lifestyle changes and market trends.

Virtual tours are here to stay, and their integration with AI will only deepen as technology evolves. From streamlining the property search process to opening up new avenues for marketing and buyer engagement, virtual tours are a testament to how AI can enhance everyday lives. For anyone interested in understanding how AI can simplify intricate, often stressful processes, virtual tours provide a clear and impactful example. The blend of convenience, efficiency, and engagement offered by this innovation truly sets a benchmark for AI applications across various sectors.

CHAPTER 13:
AI IN TRAVEL AND TOURISM

In the ever-evolving landscape of travel and tourism, AI dramatically reshapes how we plan, experience, and even conceptualize our journeys. Gone are the days of tedious travel research; personalized travel planning through AI algorithms now offers tailored itineraries based on our preferences, from flight bookings to the finest dining experiences. AI chatbots and virtual assistants are transforming customer service, providing 24/7 support, real-time translations, and instant answers to travel queries, thereby enhancing user satisfaction and reducing wait times. Furthermore, smart travel apps leverage AI to give us real-time updates on weather, traffic, and local attractions, ensuring a seamless and enriched travel experience. The future of travel isn't just about getting from point A to point B; it's about immersing ourselves in the journey itself, with AI as our ingenious guide.

Personalized Travel Planning

Travel planning, once riddled with countless hours of research and endless decision-making, has entered a new era with artificial intelligence (AI). Personalized travel planning leverages AI to curate itineraries tailored to individual preferences, making the process both efficient and enjoyable. The days of generalized travel suggestions are giving way to highly customized experiences, designed to meet the unique tastes and needs of each traveler.

Imagine planning a trip where every aspect, from flights and accommodations to activities and dining options, is optimized for you without lifting a finger. AI makes this possible by analyzing vast datasets of travel information and user preferences. This smart approach not only saves time but also ensures that each component of the trip aligns perfectly with your interests and budget.

One significant advantage of AI in travel planning is its ability to learn from your past behaviors. For instance, if you frequently book luxury hotels and favor adventurous activities, AI algorithms recognize these patterns and tailor future travel suggestions accordingly. This continuous learning loop guarantees that the recommendations you receive are increasingly relevant and personalized over time.

The integration of AI into travel planning begins with data collection. By accessing your browsing history, social media activity, and even purchase patterns, AI systems build a comprehensive profile. This profile serves as the foundation upon which personalized travel itineraries are constructed. Whether it's romantic getaways or family vacations, AI ensures that every detail resonates with your unique preferences.

Recommendation engines are at the heart of AI-powered travel planning. These sophisticated tools use machine learning to analyze data from numerous sources, including travel reviews, destination popularity, weather forecasts, and local events. By processing this information in real-time, AI can recommend the best times to visit specific locations, suggest hidden gems off the beaten path, and forecast travel trends to help you stay ahead of the curve.

One exemplary use case is the booking of flights and accommodations. AI can compare millions of options in seconds, taking into account variables such as price, convenience, and user reviews. Not only does this save you the hassle of comparison shopping, but it also ensures that you get the best deals available. Additionally, AI-powered chatbots are often integrated into travel booking platforms, offering instant assistance and answering any queries you might have.

Accommodations are another area where AI excels in providing personalized experiences. Using natural language processing (NLP), AI can parse through customer reviews to identify aspects of a hotel that appeal specifically to you, such as room amenities, customer service quality, or proximity to attractions. This granular level of detail makes it easier to find a place that aligns perfectly with your expectations.

When it comes to activities and entertainment, AI has the capability to provide curated recommendations that align with your interests. For instance, if you're an art lover visiting Paris, AI can guide you to lesser-known galleries and exclusive exhibitions that match your artistic tastes. Similarly, culinary enthusiasts can receive personalized dining suggestions based on their historical preferences and dietary needs.

Moreover, AI doesn't stop at planning; it continues to enhance your travel experience in real-time. Through mobile applications, AI can offer dynamic updates and suggestions based on your location, time of day, and even your mood. For example, if you're feeling spontaneous, the app might recommend a nearby street festival or a pop-up eatery that wasn't on your original itinerary.

Virtual tour guides powered by AI represent another leap forward in travel planning. These guides can provide immediate, contextually relevant information about landmarks, museums, and cultural sites as you explore. AI-driven voice assistants can narrate histories, explain cultural nuances, and even suggest the best vantage points for photography—all while you're experiencing the location firsthand.

AI also enhances group travel by accommodating the diverse preferences of each group member. Algorithms can analyze individual profiles to suggest activities that cater to common interests, ensuring everyone has a memorable experience without compromising on personal tastes. This is particularly useful for family vacations or corporate re-treats where varying expectations often make planning a challenge.

Travel safety is another area where AI proves invaluable. With re-al-time data analysis, AI can assess safety conditions in your destina-tion, taking into account factors like weather events, political climate, and health advisories. This proactive approach allows travelers to make informed decisions and avoid potential risks, thereby ensuring a wor-ry-free journey.

Budget management is always a concern when planning a trip, and AI provides solutions here as well. By tracking your spending patterns and comparing prices across different platforms, AI can offer dynamic budgeting advice, helping you allocate funds effectively without sacri-ficing quality. Moreover, AI can alert you to flash sales and exclusive deals, enabling you to capitalize on last-minute savings.

In a world where sustainability is increasingly important, AI can help travelers make eco-friendly choices. By analyzing the environ-mental impact of various accommodations, transportation options, and activities, AI can guide you towards greener alternatives. This en-sures that your travel plans align not only with your personal prefer-ences but also with global efforts to minimize carbon footprints.

Artificial intelligence has transformed travel planning from a com-plex, time-consuming task into a streamlined, highly enjoyable experi-ence. As you embrace AI-driven tools and applications, you open the door to a world where every trip is a perfect blend of excitement, relax-ation, and personalized touch.

There's no denying that the future of travel is smart. The continual advancements in AI technology promise even more innovative solu-tions that will make personalized travel planning an indispensable part of our lives. As you navigate this new landscape, remember that the ultimate goal of AI in travel and tourism is to enrich your experiences, making each journey a unique adventure tailored just for you.

AI in Customer Service

The travel and tourism industry is one of the sectors that has seen significant transformation with the implementation of artificial intelligence (AI), especially in customer service. From booking flights and hotels to handling in-destination queries, AI-driven customer service is enhancing convenience and elevating the overall travel experience.

AI in customer service can be seen in various forms, ranging from chatbots that provide instant responses to traveler queries to virtual assistants that offer personalized recommendations. These systems utilize natural language processing (NLP) to understand and respond to customers in a human-like manner. This technology not only reduces wait times but also ensures that travelers receive accurate and contextual information, enhancing their satisfaction.

For instance, the use of AI chatbots in answering customer inquiries is becoming increasingly popular. These chatbots are programmed to handle a wide array of questions, from the mundane to the complex. They can provide information about flight statuses, boarding gates, baggage policies, and even offer destination guides. Because they are available 24/7, they help maintain a continuous customer service experience, regardless of time zones or holidays.

The integration of AI in customer service doesn't stop at chatbots. AI-powered virtual assistants are taking personalization to the next level. By analyzing customer data, such as previous travel history, preferences, and behaviors, these virtual assistants can offer tailored recommendations. Whether it's suggesting a flight upgrade, recommending a hotel that matches the customer's past preferences, or even curating a list of local attractions, virtual assistants make the travel planning process seamless and more enjoyable.

Another significant advantage of AI in customer service is its ability to handle high volumes of inquiries efficiently. During peak travel seasons, customer service representatives can become overwhelmed with the influx of questions and issues. AI systems alleviate this pressure by managing routine inquiries and allowing human staff to focus on more complex problems. This synergy not only improves efficiency but also ensures that customer service teams are better equipped to provide high-quality assistance.

Moreover, AI-powered systems are continuously learning and evolving. Machine learning algorithms analyze vast amounts of data to understand trends and patterns, enabling these systems to improve their responses over time. The more interactions they handle, the smarter they become, resulting in progressively better service quality and customer satisfaction.

In addition to enhancing the customer experience, AI in customer service significantly reduces operational costs for travel companies. Deploying AI solutions means fewer human agents are needed to handle customer inquiries. This reduction in labor costs can be redirected into other areas, such as improving the quality of services or expanding offerings, ultimately benefiting both the company and its customers.

AI also plays a crucial role in managing crises and unforeseen events. Natural disasters, flight cancellations, and other disruptions can wreak havoc on travel plans. AI-driven customer service systems can swiftly analyze the situation and provide real-time updates and solutions to affected travelers. For example, in the event of a flight cancellation, an AI system could automatically rebook flights and notify passengers, significantly reducing the stress associated with such disruptions.

In conclusion, AI in customer service within the travel and tourism industry is revolutionizing the way businesses interact with their customers. Through chatbots, virtual assistants, and machine learning algorithms, AI is making travel planning more personalized, efficient, and enjoyable. The continuous evolution of these technologies promises even more significant improvements, ensuring that the future of travel is as smooth and seamless as possible.

As AI technology advances, its role in customer service will only become more integral. The travel industry must continue to invest in and adapt to these innovations to meet the rising expectations of tech-savvy travelers. By doing so, they can ensure a competitive edge in an increasingly digital marketplace while delivering unparalleled customer experiences.

Smart Travel Apps

In the intricate web of travel and tourism, navigating the maze of choices, schedules, and opportunities has always posed a daunting challenge. Enter smart travel apps, a revolutionary application of artificial intelligence transforming this industry by personalizing and simplifying the travel experience.

Imagine planning a vacation where every detail aligns perfectly with your interests, preferences, and even mood. Smart travel apps build on AI's advanced learning algorithms to offer deeply personalized itineraries. They consider past travel history, current market trends, user preferences, and real-time data to curate experiences that are not just convenient but also deeply satisfying. This level of personalization extends from flight bookings to restaurant recommendations, making the entire travel process smoother and more enjoyable.

Navigation is another domain where smart travel apps shine. AI-powered mapping tools provide more than just the shortest route to your destination. They factor in live traffic conditions, weather forecasts, and local events to suggest optimal routes. This real-time information helps travelers avoid delays and maximize their time. Moreover, these apps can offer offline capabilities, ensuring that users can access crucial information even without internet connectivity.

Language barriers have long been a formidable obstacle for travelers. AI-driven translation features within travel apps can now offer near-instantaneous translations, allowing users to communicate effortlessly, whether they are in Tokyo or Timbuktu. These translations are not limited to mere text; they can also include voice-to-voice translation, helping navigate conversations with locals with ease. This technological leap fosters better cultural understanding and more meaningful interactions.

Accommodation is another key area where smart travel apps make a significant impact. AI systems can sift through thousands of options, taking into account not only price and location but also user reviews, amenities, and even the ambiance of the neighborhood. By analyzing this extensive array of data, the apps can recommend the best possible lodging tailored to the traveler's specific needs, whether they're looking for a cozy bed and breakfast or a luxurious international hotel chain.

Safety and security, integral components of travel, have been enhanced considerably through AI. Many smart travel apps come integrated with features that alert users to unsafe areas, medical facilities in the vicinity, and emergency contact numbers. AI-driven risk analysis tools can offer real-time updates on political instability, natural disasters, or health risks in potential travel destinations, helping travelers make informed decisions.

Another fascinating application of AI in travel apps is predictive pricing. By leveraging massive amounts of historical data, these apps can predict the best times to book flights and accommodations to get the best deals. This ability to foresee price changes lets travelers optimize their budgets without compromising on their dream trips.

Beyond just planning and booking, AI enhances the on-the-ground travel experience. Augmented reality (AR) features in smart travel apps can help users explore their surroundings in a more immersive way. For instance, by pointing their phone at a historical monument, users can receive rich, interactive information about its history, significance, and hidden secrets.

Customer service in travel has also seen a paradigm shift thanks to AI. Virtual assistants and chatbots provide 24/7 support, answering queries, resolving issues, and giving tailored recommendations. These AI-driven solutions can handle a wide range of tasks, from rebooking cancelled flights to providing dining recommendations, ensuring that travelers receive timely and efficient support throughout their journey.

For frequent travelers, loyalty programs present another layer of complexity that AI simplifies elegantly. Smart travel apps track user preferences and engagement levels with different brands and services, optimizing loyalty points and rewards. These systems ensure users get the best possible benefits, often automating the application of discounts and special offers.

Moreover, AI could play a crucial role in eco-friendly travel. Smart travel apps can recommend sustainable options, such as eco-friendly lodging, carpooling services, or restaurants that practice waste reduction. By making sustainability accessible and convenient, these applications encourage more travelers to opt for greener choices, contributing to global efforts to protect the environment.

The future potential for smart travel apps is almost limitless. As AI technologies continue to evolve, we can expect even more sophisticated capabilities integrated into these platforms, making travel not only more accessible but also more enriching. Imagine personalized travel itineraries generated through facial recognition technologies assessing your emotional state, or even immersive travel experiences curated by advanced AI based on your biometrics and mood.

Smart travel apps already illuminate the power and practicality of AI in everyday life. They exemplify a transformative shift toward seamless, personalized, and enriched travel experiences, reflecting AI's potential to streamline and enhance our daily routines beyond mere convenience. As we continue to explore and develop these technologies, the way we approach travel could become as effortless as tapping a screen, opening up a new world of possibilities for adventurers and casual explorers alike.

CHAPTER 14:
AI IN FITNESS AND WELLNESS

Artificial Intelligence is revolutionizing the fitness and wellness industry, bringing a new dimension to personal health and well-being. Imagine a world where your workouts are not only customized to your unique needs but also adaptive based on real-time data collected from smart wearables. AI-driven platforms create personalized workout plans that evolve with your progress, ensuring you're always challenged and motivated. Beyond physical fitness, AI technologies extend their benefits to mental health by offering tools for meditation, mindfulness, and stress management, tailored to individual emotional states. These intelligent systems act as 24/7 wellness coaches, helping you maintain a balanced and healthy lifestyle, making wellness accessible and achievable for everyone. By seamlessly integrating into everyday routines, AI is turning the futuristic ideal of holistic health into a tangible reality for many.

Personalized Workout Plans

Imagine having a personal trainer, nutritionist, and health coach all rolled into one and available 24/7, tailored just for you. This isn't a distant dream but a present reality thanks to the integration of AI in fitness and wellness. Personalized workout plans driven by AI are revolutionizing the way we approach exercise by providing highly customized, data-driven fitness regimens that adapt to our unique needs, goals, and daily performance.

The cornerstone of these AI-driven plans is their ability to leverage massive amounts of data. Various fitness trackers and smart wearables collect detailed information about our activities, from steps taken to heart rate variability, sleep patterns, and even stress levels. By analyzing this data, AI algorithms can identify patterns and provide insights into our health and fitness that would be impossible for a human trainer to discern. This allows for the creation of workout plans that aren't just personalized but dynamically responsive to your body's signals.

One of the most significant advantages of AI-driven personalized workout plans is their adaptability. Traditional workout plans often follow a one-size-fits-all approach that may not account for individual differences in fitness levels, recovery times, or physical limitations. In contrast, AI can continuously adjust the workout intensity, type, and duration based on your ongoing performance, ensuring that you're always training efficiently and safely. If you're fatigued, the AI might suggest a lighter session focusing on recovery. If you're progressing faster than anticipated, it could ramp up the intensity to keep you challenged.

A personalized AI workout plan often starts with an initial assessment to understand your current fitness level, goals, and any specific needs or constraints you may have. This might include a series of fitness tests, a review of your medical history, and a questionnaire on your exercise preferences and lifestyle. The AI then crafts a plan that aligns with your objectives, whether it's weight loss, muscle gain, improved endurance, or general health and wellness. What sets AI apart is that this plan is a living document; it evolves as you do, using real-time data to make informed adjustments.

Moreover, AI can offer more than just physical training plans. It can incorporate recommendations for nutrition, hydration, and even mental health practices. For instance, if your sleep data from a wearable device shows you're not getting enough rest, the AI might suggest adjustments not only to your workout schedule but also provide tips on improving sleep hygiene. In this way, it offers a holistic approach to health and fitness.

Interaction with AI-driven workout plans can be surprisingly human-like. Thanks to advancements in natural language processing (NLP), these systems can communicate in a way that feels intuitive and personalized. You might receive encouraging messages, reminders, and feedback that help keep you motivated. Some applications even allow you to ask questions or discuss your progress, making the experience highly interactive and engaging.

The social aspect of fitness isn't overlooked either. Many AI platforms offer community features where you can connect with other users following similar plans. These communities provide additional motivation, support, and a sense of accountability—key factors in staying committed to a fitness journey. Additionally, AI can help you set up challenges, track your progress relative to peers, and even organize virtual training sessions, making the often solitary pursuit of fitness a more communal experience.

One might wonder about the accuracy and safety of these AI-generated plans. It's essential to note that many of these systems are developed in partnership with fitness experts, sports scientists, and medical professionals, ensuring that the AI's recommendations are grounded in scientific principles and best practices. Furthermore, continuous data monitoring helps the AI to detect any potential issues early, prompting you to seek professional advice if necessary.

Also, the integration of AI in fitness opens up new possibilities for accessibility and inclusivity. Personalized AI workout plans can be customized for individuals with disabilities, chronic illnesses, or other conditions requiring special attention. This means that fitness and wellness are within reach for everyone, not just those who can afford personal trainers or have access to specialized gyms. The democratization of fitness expertise through AI can lead to a healthier, more inclusive society.

In terms of motivation, AI offers unique features that help keep you on track. Gamification elements, such as earning points, unlocking achievements, and competing in leaderboards, add a fun and engaging dimension to your workouts. The AI also provides insightful feedback on your progress, highlighting milestones and celebrating achievements, no matter how small. These positive reinforcements play a crucial role in maintaining motivation and commitment over the long term.

A case in point is the rise of smart home gyms, which combine advanced AI with exercise equipment to offer a truly immersive experience. These setups often include real-time form correction, automated resistance adjustments, and virtual classes guided by expert trainers. By blending technology with fitness equipment, AI takes workout personalization to a whole new level, providing an experience that closely mirrors and, in some cases, even surpasses what you might get in a traditional gym setting.

Looking ahead, the potential for AI in personalized workout plans is immense. With continual advancements in AI technology, including more refined algorithms and better integration of biometric data, the personalization and effectiveness of these plans will only improve. Future developments might include even more sophisticated ways to simulate personal training experiences or offer predictive insights that preempt injuries or health issues before they arise.

In conclusion, AI's role in crafting personalized workout plans symbolizes a significant leap forward in the realm of fitness and wellness. It brings a new level of precision, adaptability, and accessibility to workout planning that was previously unimaginable. By leveraging the power of data and intelligent algorithms, AI can transform your fitness journey, offering a personalized blueprint to achieving your health and wellness goals. Embracing these technologies can pave the way for a more active, balanced, and healthier lifestyle.

AI in Mental Health

The integration of Artificial Intelligence (AI) into mental health is revolutionizing the way we approach emotional well-being. At the intersection of technology and psychology, AI is providing unprecedented opportunities for diagnosis, treatment, and support. With a growing recognition of mental health's importance, AI-driven tools are transforming this critical area by making therapeutic resources more accessible, personalized, and consistent.

One of the most impactful applications of AI in mental health is in diagnosis and monitoring. Traditional methods often rely heavily on self-reporting and clinical interviews, which can be subjective and inconsistent. AI, however, can analyze a multitude of data points to identify patterns indicating mental health issues. For instance, natural language processing (NLP) algorithms can examine speech and text for indicators of depression, anxiety, or other disorders. These systems can flag concerning phrases or changes in communication styles long before a human might notice, offering a proactive approach to mental health care.

Additionally, AI's power lies in its ability to process enormous amounts of data. This data can come from sources like electronic health records, wearable devices, and even social media activity. Machine learning models can then identify trends and potential triggers for mental health concerns, providing a more holistic and accurate picture of an individual's mental state. Such insights are invaluable for clinicians, who can tailor interventions more precisely and effectively.

Moreover, AI is democratizing access to mental health support. Chatbots and virtual therapists are increasingly sophisticated, offering immediate, 24/7 assistance to those in need. These AI-driven tools provide a range of services, from cognitive-behavioral therapy (CBT) techniques to mindfulness exercises. They serve not as a replacement for human therapists but as a supplementary resource, especially beneficial for those who may face barriers to accessing traditional mental health services, such as geographic limitations or stigma.

For instance, Woebot, an AI chatbot developed by psychologists and AI experts, uses NLP to engage users in conversations about their mental health. It provides evidence-based techniques derived from CBT to help users manage their emotions and develop healthier thought patterns. Its accessibility and user-friendly nature make it a valuable tool for early intervention and support.

Furthermore, AI can enhance the therapeutic process even when human therapists are involved. Tools like video analysis software can assess patient-therapist interactions, providing feedback that can help improve communication and therapeutic outcomes. AI can also assist in developing personalized treatment plans by analyzing which interventions have been most effective for similar patients, ensuring the treatment is tailored to individual needs.

Another exciting development is the use of AI in predicting and preventing crises. Machine learning models can analyze data to forecast when someone might be at risk of a mental health crisis, such as a panic attack or suicidal ideation. These predictive models can alert healthcare providers or caregivers, enabling timely interventions that could save lives.

Additionally, AI's capability in mental health extends to the realm of neuroscience. Techniques like Functional Magnetic Resonance Imaging (fMRI) and Electroencephalography (EEG) produce vast amounts of data about brain activity. AI is instrumental in processing and interpreting this data, identifying neural patterns associated with various mental health conditions. This understanding can lead to more effective treatments, potentially even curing some conditions that were previously deemed untreatable.

Despite the promising developments, it's essential to acknowledge and address challenges and ethical considerations related to AI in mental health. Privacy concerns are particularly pertinent, as mental health data is highly sensitive. Developers and healthcare providers must ensure robust data protection protocols and obtain informed consent from users. Furthermore, there's the issue of algorithmic bias. AI models are only as good as the data they're trained on. If the data is biased, the diagnostic and treatment recommendations could perpetuate existing inequalities.

Moreover, the human touch in mental health care remains irreplaceable. AI can augment but not replace the empathy and understanding that human therapists provide. The goal should be to use AI in a way that enhances human capabilities and improves patient outcomes, not to create a cold, impersonal healthcare experience.

We also need to consider the digital divide. Not everyone has access to the necessary technology, such as smartphones or reliable internet connections, to benefit from AI-driven mental health tools. This could exacerbate existing disparities in mental health care access.

Another aspect to consider is the mental health of the professionals themselves. AI can help reduce the workload and administrative burden on mental health practitioners, allowing them to focus more on patient care. Tools that automate scheduling, patient follow-ups, and documentation can free up valuable time, reducing burnout and improving the overall quality of care.

In education and training, AI is also making strides. Virtual reality (VR) and augmented reality (AR) platforms powered by AI can simulate real-life scenarios for training therapists. These platforms can provide instant feedback and help trainees develop their skills in a controlled, risk-free environment.

In conclusion, AI stands at the forefront of a mental health revolution. Its capabilities in diagnosis, treatment, and ongoing support offer a beacon of hope for millions. By breaking down barriers to access, providing personalized care, and enabling early interventions, AI is set to significantly enhance mental health outcomes. However, it is crucial to navigate the ethical landscapes carefully, ensuring this technology benefits everyone equally, while maintaining the irreplaceable human elements of empathy and personal connection in mental health care.

Smart Wearables

In today's fast-paced world, staying fit and maintaining overall wellness can sometimes feel like an uphill battle. This is where smart wearables, powered by sophisticated AI algorithms, come in, acting as personal assistants that help you stay on top of your health and fitness goals. These devices, ranging from smartwatches to fitness trackers, have become increasingly popular due to their ability to provide real-time data and actionable insights.

One of the most transformative aspects of smart wearables is their capability to monitor physical activity and vital signs. Whether you're running a marathon or simply taking a walk, these devices record your steps, heart rate, and calories burned, offering a comprehensive overview of your daily activity. AI algorithms analyze this data to provide customized feedback, enabling you to make informed decisions about your fitness regimen. No longer do you need to guess whether you're on track; the evidence is right there on your wrist.

But smart wearables do more than just track physical activity. They play a crucial role in managing overall wellness by monitoring sleep patterns, stress levels, and even providing reminders to take a break or move around. Poor sleep can significantly affect your well-being, and these devices can help you understand the quality of your sleep cycles and suggest improvements. The integration of AI allows for the interpretation of complex data sets, giving you personalized recommendations for optimizing rest and stress management.

Moreover, smart wearables are evolving rapidly, with new features being added all the time. For example, some devices now include ECG (Electrocardiogram) functionality to monitor heart health more closely. These capabilities were once confined to clinical settings but are now available right at your fingertips. AI algorithms analyze the collected ECG data to detect irregular heartbeats or other anomalies, offering an early warning system that could potentially save lives. The immediacy of feedback is one of the most significant benefits, making healthcare more proactive rather than reactive.

Advanced AI technologies also enable these wearables to offer personalized workout plans. Depending on your fitness level, goals, and even your daily mood, the device can suggest specific exercises and routines. It's almost like having a personal trainer on call 24/7. Whether you're aiming to lose weight, build muscle, or maintain general fitness, AI ensures that your workout plan is as effective and enjoyable as possible, tailoring suggestions to fit your lifestyle and preferences.

Another interesting application of AI in smart wearables is nutritional advice. Some advanced devices can track your dietary habits and offer personalized meal suggestions based on your goals and nutritional needs. By integrating with other apps and platforms, these wearables can create a holistic profile of your lifestyle, helping you make better dietary choices. Whether you're looking to count calories, balance macronutrients, or adhere to a specific diet, AI-driven insights make it easier than ever to stay on track.

What makes these devices indispensable is their ability to provide real-time feedback and updates. Imagine running a few miles and receiving instant updates on your pace, heart rate, and hydration levels.

Such real-time data can make each workout session more effective by offering insights on how to tweak your form or adjust your pace to achieve better results. For those training for specific goals, like a marathon or a triathlon, this instant feedback loop can be a game-changer.

Beyond physical health, smart wearables are making significant strides in mental wellness. Features like guided meditation, breathing exercises, and even mood tracking have become increasingly common. By analyzing data on your daily routines, activity levels, and even subtle indicators like voice tone or facial expressions, AI can offer recommendations to improve mental health. This could entail suggesting a moment of mindfulness in the middle of a hectic day or a breathing exercise to manage stress.

Safety is another area where smart wearables shine. Features like fall detection, emergency SOS, and real-time location tracking can be life-savers, especially for the elderly or people with medical conditions. By employing machine learning algorithms, these devices can discern regular activities from potentially dangerous situations, triggering alerts and sending real-time data to emergency contacts or health services.

However, the potential of smart wearables doesn't end with just individual health and fitness. These devices are also contributing to broader health trends and research. Aggregated, anonymized data from millions of users provide invaluable insights for medical research, public health policies, and even urban planning. The vast amount of data collected through smart wearables serves as a goldmine for researchers, enabling them to identify trends, predict outbreaks, and understand long-term health impacts like never before.

The interconnectivity of smart wearables with other smart devices also can't be overlooked. Integration with smartphones, smart home devices, and even vehicles allows for a more seamless experience. Your wearable can remind you to turn off the lights when you're about to sleep or suggest optimal light settings for better sleep quality. The possibilities are almost endless, creating an ecosystem where everything works in harmony to improve the quality of your life.

Despite all these advancements, the journey of smart wearables is still just beginning. Upcoming innovations are likely to include even more sophisticated biometric sensors, improved battery life, and further miniaturization of hardware. The future might bring us wearables that can continuously monitor blood glucose levels, detect a broader range of biomarkers, and even predict potential health issues before they become problems.

Of course, the ethical considerations and data privacy issues surrounding smart wearables cannot be ignored. As these devices collect an ever-increasing amount of personal and sensitive data, ensuring that this information is securely stored and responsibly used becomes crucial. Users must have control over their data, knowing exactly what is being collected and how it is being used. As we advance, robust AI governance frameworks will be essential to maintaining trust and maximizing the benefits of these remarkable devices.

In conclusion, smart wearables are revolutionizing the landscape of fitness and wellness. By leveraging the power of AI, these devices offer personalized, real-time insights that empower users to make informed decisions about their health. From tracking physical activity and monitoring vital signs to enhancing mental wellness and contributing to broader health research, the potential applications are vast and varied. As technology continues to evolve, smart wearables will undoubtedly become even more integral to our daily routines, enhancing our overall well-being in ways we are only beginning to imagine.

CHAPTER 15:
AI IN AGRICULTURE

In the ever-evolving field of agriculture, Artificial Intelligence is emerging as a game-changer, bringing revolutionary changes that promise to enhance efficiency and yield. By leveraging AI-powered technologies, farmers can now utilize precision farming techniques to optimize resource usage such as water, fertilizers, and pesticides, ultimately leading to healthier crops and higher productivity. Advanced AI algorithms in crop monitoring systems provide real-time data analysis and predictive insights, enabling timely decisions and interventions to mitigate risks like disease outbreaks and climate variability. Additionally, automated farming equipment, including robots and drones, can perform labor-intensive tasks such as planting, harvesting, and pest control with unparalleled accuracy and speed. These innovations not only reduce manual labor but also contribute to sustainable farming practices, highlighting AI's potential to revolutionize agriculture and secure food resources for the future.

Precision Farming

Precision farming, also known as precision agriculture, represents a transformative evolution in agricultural practices, driven by advancements in artificial intelligence. The core idea is to use data to guide farming decisions with higher accuracy and efficiency, ultimately leading to enhanced productivity and resource management. Far from the traditional, often intuitive-based methods, precision farming leverages technology to deliver empirical data that helps farmers make more informed decisions. This shift holds promise not only for increasing crop yields but also for reducing environmental impacts.

At the heart of precision farming is the collection and analysis of data from various sources. These sources include remote sensing technologies, such as drones and satellites, ground-level sensors, and even the Internet of Things (IoT) devices. This data provides insights into soil conditions, crop health, weather patterns, and pest populations. By collating and analyzing this data, AI-driven systems can provide actionable recommendations tailored to specific fields, crops, and even individual plants.

Let's delve into how this data collection and analysis works in practice. Imagine a farmer managing a large field with varied soil conditions. Traditionally, the farmer might apply a uniform amount of fertilizers and pesticides across the entire field. This could lead to some areas receiving excess chemicals, while others might not get enough. With precision farming, sensors embedded in the soil and aerial imagery can create a detailed map of soil fertility and plant health. AI algorithms analyze this data to recommend variable rate applications of inputs, ensuring each part of the field gets exactly what it needs. This not only optimizes growth but also minimizes waste and runoff, thereby protecting the environment.

One significant aspect of precision farming is *variable rate technology* (VRT). VRT systems use GPS-guided equipment to vary the rate of inputs like seeds, fertilizers, and pesticides. AI algorithms determine the optimal rates for different sections of the field, enhancing efficiency and crop yield while limiting environmental impact. Farmers can set these systems to either react in real-time to sensor data or follow prescriptive maps generated by AI analysis.

Another critical component is the role of drones and satellite imagery. High-resolution images captured from these devices enable farmers to monitor crop health, growth patterns, and even detect early signs of diseases or pest infestations. AI algorithms can analyze these images to identify areas that need attention, allowing for timely interventions. For instance, machine learning models can differentiate between healthy and stressed plants by analyzing color variations and patterns. This level of monitoring was previously unattainable with traditional farming techniques.

Beyond individual farms, precision farming has potential benefits on a larger scale, such as regional or even national food security. Aggregated data from numerous farms can give insights into crop performance trends, helping governments and organizations to better anticipate food supply needs and respond to potential shortages. For instance, AI can predict the likely impact of adverse weather events on crop yields and recommend strategies to mitigate these risks.

Additionally, AI-powered predictive analytics play a pivotal role in precision farming. Predictive models can forecast weather conditions, pest outbreaks, and crop performance based on historical data and real-time inputs. These predictions enable farmers to take preemptive actions, such as adjusting irrigation schedules before a drought or applying pest control measures before an outbreak becomes severe. This proactive approach helps ensure consistent crop quality and yield.

Precision farming also promotes sustainable agriculture, which is crucial in a world where resource conservation is increasingly important. Optimizing the use of water, fertilizers, and pesticides reduces the ecological footprint of farming activities. AI can help manage water usage through smart irrigation systems that monitor soil moisture levels and weather forecasts to apply water only when and where needed. This not only conserves water but also improves crop health by preventing overwatering.

Economic benefits are another significant driver for adopting precision farming technologies. By precisely managing inputs, farmers can reduce costs associated with fertilizers, pesticides, and water. Improved crop yields and quality can lead to higher market prices and increased profitability. Though the initial investment in technology and equipment can be substantial, the return on investment often justifies the expenditure through long-term gains.

Despite its many advantages, the adoption of precision farming is not without challenges. One major hurdle is the high cost of advanced technology and equipment, which can be prohibitive for small-scale farmers. Additionally, there is a learning curve associated with using these technologies effectively. Access to reliable internet connectivity and data infrastructure is also crucial, which can be a limitation in rural areas. Even so, as the technology becomes more affordable and accessible, these barriers are gradually being overcome.

Human expertise and judgment remain indispensable in precision farming. While AI can process vast amounts of data and offer recommendations, farmers' experience and intuition are still crucial in making final decisions. The optimal use of precision farming involves a harmonious blend of technology and traditional knowledge, leveraging the strengths of both to achieve the best outcomes.

In conclusion, precision farming embodies the spirit of innovation by effectively integrating AI into agriculture. As we look toward the future, the continued development and adoption of these technologies promise to revolutionize farming practices. By harnessing the power of data and AI, farmers can cultivate not just crops, but a more sustainable and prosperous future for us all.

AI in Crop Monitoring

Imagine a world where farmers can predict how their crops will perform with the same precision a meteorologist predicts tomorrow's weather. Thanks to AI in crop monitoring, that world is closer to reality than you might think. AI has emerged as a powerful ally to the agriculture sector, helping farmers make data-driven decisions that maximize yield and minimize waste.

Crop monitoring is a multi-faceted process involving the collection of data from various sources such as satellite imagery, drones, soil sensors, and weather stations. These data points are then analyzed using AI algorithms to give farmers real-time insights into the health and progress of their crops. It's a game-changer, shifting farming from a labor-intensive occupation to a high-tech industry.

One of the primary benefits of AI in crop monitoring is its ability to detect disease early. For instance, computer vision technology can examine plant images to identify signs of disease that might be invisible to the human eye. By detecting these signs early, farmers can take preventative measures before the disease spreads, thus saving their crops and reducing the need for chemical interventions. This is not only economically advantageous but also environmentally friendly.

Beyond disease detection, AI-powered crop monitoring systems can also assess soil health by analyzing data from soil sensors. These sensors measure various parameters like moisture levels, pH, and nutrient content. Using machine learning algorithms, the gathered data provides actionable insights into soil conditions, suggesting when and where to water or fertilize. This precise management ensures optimal growth conditions and reduces resource wastage.

Weather is another critical aspect of farming that AI has a hand in managing. Predictive analytics, powered by AI, can analyze historical weather data and current conditions to forecast future weather events. This allows farmers to make informed decisions about planting, harvesting, and other critical activities. Accurate weather predictions can help avoid potential damage from unexpected weather events, ensuring that crops are protected and resources efficiently utilized.

AI also plays a significant role in pest management. Traditional methods often involve blanket application of pesticides, which can be harmful to the environment. AI systems, however, can analyze data from various sources to predict pest outbreaks and recommend targeted actions. This approach not only reduces the amount of pesticide used but also minimizes the environmental impact.

Implementing an AI-based crop monitoring system may seem like a significant investment, but its long-term benefits far outweigh the initial costs. The efficiency and accuracy that AI brings to crop monitoring can lead to increased profitability and sustainability for farms of all sizes. Moreover, these technologies are becoming increasingly accessible, thanks to advancements in hardware and software development.

One of the most exciting developments in AI for crop monitoring is the integration of drone technology. Drones equipped with high-resolution cameras and sensors can cover large areas of farmland quickly, capturing detailed images and data. These aerial views offer an unparalleled perspective of crop health, allowing for early detection of problems such as nutrient deficiencies or weed infestations. When combined with AI, drones provide real-time, actionable insights that can substantially improve farming outcomes.

Data is the backbone of AI in crop monitoring. Without accurate data, even the most sophisticated AI algorithms will flounder. Data collection needs to be robust and continuous, incorporating information from various sources, including satellite imagery, ground-based sensors, and historical farming records. The challenge lies in not just collecting this data but ensuring that it is clean, reliable, and relevant.

Farmers, however, don't need to be experts in AI to benefit from these technologies. Many modern crop monitoring solutions are designed with user-friendly interfaces that translate complex data and algorithms into easily understandable insights. These systems often come with mobile apps that send real-time alerts, ensuring farmers are always informed and can act promptly.

There's also an exciting intersection between AI and IoT (Internet of Things) in the realm of crop monitoring. IoT devices such as smart sensors, weather stations, and automated irrigation systems can connect to AI platforms, creating an integrated ecosystem that brings a holistic approach to farm management. This combination provides a seamless flow of information, enabling more precise and timely interventions.

Precision farming is another term closely associated with AI in crop monitoring. While traditional farming methods can be quite generalized, precision farming, powered by AI, focuses on optimizing every farming activity down to the smallest detail. This includes pinpointing the exact amount of water needed for each plant, the precise time to harvest each crop, and the optimal use of fertilizers and pesticides. The result is a more efficient and sustainable farming system that maximizes yield with minimal resource input.

AI's contribution to crop monitoring doesn't stop at the analysis. Predictive models can suggest customized treatment plans for different crop issues. These models can be tailored to account for the specific type of crop, its growth stage, and even local environmental conditions. Such personalized solutions ensure that farms are not subject to one-size-fits-all recommendations but receive advice that is uniquely suited to their specific needs.

While the benefits of AI in crop monitoring are evident, there are challenges to widespread adoption. Limited access to technology, high costs, and insufficient knowledge about AI are common barriers, especially for smaller farms. However, ongoing efforts to educate farmers and lower the cost of technology are gradually addressing these issues.

The role of AI in crop monitoring resonates far beyond individual farm boundaries. By enhancing productivity and sustainability, it can contribute to global food security. With the world's population continuing to grow, the pressure on agricultural systems to produce more food with fewer resources is immense. AI in crop monitoring can be a crucial part of meeting this challenge.

In conclusion, AI in crop monitoring represents a significant leap forward for the agriculture sector. From early disease detection to precise soil management, weather forecasting, and pest control, AI-powered systems offer comprehensive solutions that make farming more efficient and sustainable. As these technologies continue to evolve and become more accessible, their impact on agriculture – and by extension, on our daily lives – will only grow. The future of farming promises to be as intelligent as it is fruitful, driven by the strides made in AI and its versatile applications in crop monitoring.

Automated Farming Equipment

In recent years, advances in AI have transformed agriculture, bringing about the emergence of automated farming equipment. These innovations are not just about easing the workload for farmers; they're revolutionizing the entire agricultural landscape. The integration of AI into farming machinery can significantly boost productivity, precision, and sustainability.

One of the most glaring advantages of automated farming equipment is its ability to operate with minimal human intervention. AI-powered tractors, for instance, can till fields, plant seeds, and apply pesticides with extreme precision. By using GPS technology combined with AI algorithms, these machines can navigate fields with pinpoint accuracy, ensuring that no section is left unattended. The result is an optimal use of resources like seeds, fertilizers, and pesticides, which leads to higher crop yields and reduced environmental impact.

Beyond tractors, there are AI-driven harvesters capable of distinguishing between ripe and unripe crops. These machines use computer vision technology to assess the condition of fruits and vegetables before picking them. As they move through the fields, they can make real-time decisions, minimizing waste and ensuring that only the best produce is harvested. Such automation guarantees that crops are picked at the perfect time, preserving maximum freshness and nutritional value.

But the role of AI in farming equipment isn't limited to planting and harvesting. Irrigation systems have also seen remarkable advancements. Smart irrigation systems equipped with sensors and AI can analyze soil moisture levels, weather forecasts, and crop requirements to determine the optimal watering schedule. This leads to water conservation and healthier crops. As water scarcity becomes a more pressing global issue, these systems offer a sustainable solution to one of agriculture's biggest challenges.

In addition to enhancing productivity, automated farming equipment can significantly reduce labor costs. Agriculture has traditionally been labor-intensive, requiring a substantial workforce for planting, maintaining, and harvesting crops. In regions where labor shortages are a concern, AI-powered machines can fill the gap, ensuring that agricultural operations continue smoothly. This is particularly relevant in the context of aging farming populations in many parts of the world.

Moreover, automated equipment can work around the clock, unaffected by fatigue or adverse weather conditions. Traditional farming methods often rely on suitable climatic windows and the physical stamina of workers. AI-enabled machinery can operate continuously, maximizing the use of available time and ensuring that time-sensitive agricultural activities are performed within the optimal periods.

Another transformative aspect of AI-driven farming equipment is its ability to collect and analyze data. These machines are often equipped with a plethora of sensors that gather information about soil health, crop conditions, pest presence, and more. By leveraging machine learning, this data can be turned into actionable insights. For example, if a specific part of a field is found to have soil deficiencies, automated equipment can address the issue directly, applying the right nutrients precisely where needed.

Furthermore, drones equipped with AI capabilities are increasingly being used for farm surveillance and crop monitoring. They can fly over fields to capture high-resolution images and videos, providing farmers with a bird's eye view of their crops. These aerial assessments allow for the early detection of issues such as pest infestations or disease outbreaks, enabling prompt intervention and minimizing damage. The drones can also assist in tasks like spraying pesticides or monitoring livestock, providing a versatile tool for modern farming needs.

While the benefits of automated farming equipment are substantial, it's important to recognize the challenges that accompany this technological shift. One significant hurdle is the initial investment required for acquiring AI-powered machinery. For small-scale farmers, the costs can be prohibitive. However, as technology advances and becomes more affordable, economies of scale may help make these tools accessible to a broader range of farmers. Governments and agricultural organizations can also play a pivotal role in offering subsidies or incentives to encourage the adoption of automated farming solutions.

Another concern is the need for adequate training and support. Farming communities must be educated on how to effectively use and maintain these sophisticated machines. This requires a concerted effort from technology providers, agricultural extension services, and educational institutions to offer training programs and resources that equip farmers with the necessary skills.

Despite these challenges, the long-term benefits of automated farming equipment are clear. By integrating AI into agricultural practices, we're paving the way for a more productive, efficient, and sustainable future. As technology continues to evolve, we'll likely see even more innovative applications of AI in farming, addressing an ever-wider range of agricultural challenges.

The potential of AI in agriculture extends beyond just solving existing problems; it promises to redefine the very nature of farming. Imagine a future where farms are entirely managed by intelligent machines, each decision and action meticulously calculated for maximum efficiency and minimal environmental impact. Such a scenario isn't just a fantasy; it's a glimpse into the transformative power of AI and its ability to reshape one of humanity's oldest and most essential industries.

Automated farming equipment represents a significant leap forward in agricultural technology. Its advantages in increasing productivity, optimizing resource use, and reducing labor costs are driving a new era of farming. As we continue to embrace and innovate upon these technologies, we move closer to a future where the world's growing food needs can be met sustainably and efficiently. The intersection of AI and agriculture holds immense promise, and we're only just beginning to scratch the surface of what's possible.

CHAPTER 16:
AI IN ENVIRONMENTAL
SUSTAINABILITY

Artificial Intelligence is emerging as a powerful ally in the quest for environmental sustainability, offering real-world solutions for some of the planet's most pressing issues. From optimizing energy consumption through smart grids to revolutionizing waste management by sorting recyclables more efficiently, AI systems can significantly reduce human impact on the environment. Environmental monitoring technologies, powered by complex algorithms, can predict natural disasters, track wildlife populations, and measure pollution levels with unprecedented accuracy. These innovations not only help conserve natural resources but also empower communities to make data-driven decisions that protect ecological balance. As societies worldwide face the immense challenge of climate change, AI acts as both a guiding hand and a powerful tool for crafting a more sustainable future, ultimately linking technological advancement with environmental stewardship.

Smart Energy Management

The rise of Artificial Intelligence (AI) in everyday life presents an astonishing opportunity to optimize energy consumption, an area that often goes unnoticed in daily routines but significantly impacts environmental sustainability. Smart energy management is one of AI's most practical and promising applications, focusing on the intelligent regulation and conservation of energy resources.

Imagine you wake up in the morning, and your house's heating system has already adjusted the temperature to your preference based on historical data and current weather conditions. This is not science fiction; it is a manifestation of AI in smart energy management. Intelligent thermostats analyze your routines and external factors to ensure optimal energy use, reducing both your utility bill and carbon footprint.

One crucial aspect of smart energy management is demand response. Employing AI algorithms, these systems can forecast energy demand and adjust energy usage patterns accordingly. For instance, during peak hours, AI can reduce the usage of non-essential appliances to prevent overloading the grid, thereby enhancing its stability. This proactive approach also opens the door to renewable energy sources such as solar and wind, which are inherently variable.

AI in smart energy management extends well beyond the confines of a single home or building. Entire smart grids leverage AI to balance energy loads across a network, optimizing for efficiency and sustainability. These grids predict energy consumption trends in real-time, allowing utility companies to make informed decisions about energy distribution. By integrating renewable energy sources, smart grids can dynamically manage the variability of these sources, ensuring a stable and reliable energy supply.

Consider the role of AI in renewable energy optimization. AI-driven models can predict solar and wind energy production with remarkable accuracy, allowing for better integration of these resources into the power grid. Advanced algorithms analyze weather patterns, historical data, and current energy demands to make these predictions. This level of precision makes it easier to rely on renewable energy, reducing our dependence on fossil fuels and mitigating climate change.

Moreover, AI doesn't just help in producing energy but also in storing it. For example, AI algorithms can manage battery storage systems more efficiently by analyzing consumption patterns and predicting future energy needs. This enables the storage systems to charge during off-peak hours and discharge when demand is high, optimizing energy use and cutting costs.

Another transformative application of AI in smart energy management is in industrial and commercial buildings. AI systems can monitor and manage energy usage across large facilities using a network of sensors and smart meters. These systems identify inefficiencies and offer actionable insights to facility managers. Whether it's adjusting the HVAC system or turning off unused machinery, AI ensures energy is consumed more efficiently, leading to significant cost savings and a reduced carbon footprint.

Transportation is another sector where smart energy management is making waves. Electric vehicles (EVs) are becoming more common, and with AI, they can be smarter in their energy use. AI helps in route optimization, reducing energy consumption by mapping out the most efficient paths. Additionally, AI can manage the charging infrastructure by determining the best times to charge based on grid demand and availability of renewable energy.

Household appliances are also joining the smart energy movement. Refrigerators, washing machines, and even lighting systems now come equipped with AI algorithms designed to optimize their energy use. For instance, a smart refrigerator can regulate its cooling cycles based on usage patterns or external temperatures. Similarly, intelligent lighting systems adjust brightness and on/off schedules based on human presence and daylight availability, ensuring energy is not wasted.

AI-based predictive maintenance is another crucial aspect of smart energy management. By analyzing data from various sensors, AI can predict when equipment is likely to fail or require servicing, enabling timely maintenance. This not only prolongs the life of the equipment but also ensures they are running at peak efficiency, saving energy and reducing operational costs.

AI also plays a significant role in energy trading and management platforms. These platforms utilize machine learning algorithms to analyze market conditions, consumer behavior, and energy consumption trends. By doing so, they can optimize the buying and selling of energy in real-time, maximizing financial returns while ensuring a sustainable and efficient energy supply.

Residential communities and business complexes are increasingly adopting AI-based microgrids. These are localized grids that can operate independently or in conjunction with the main grid. AI systems manage these microgrids by distributing energy based on real-time consumption data and predictions, thereby enhancing energy reliability and sustainability.

Another fascinating application of AI in smart energy management is in the realm of behavioral change. AI can encourage and educate individuals and organizations on energy-saving practices by providing personalized insights and recommendations. This can create a culture of energy conservation, where every small action contributes to significant overall savings.

Blockchain technology, combined with AI, offers even more potential for smart energy management. Blockchain ensures transparency and security in energy transactions, while AI optimizes these transactions for efficiency. This synergy can facilitate peer-to-peer energy trading, allowing consumers to buy and sell excess energy directly to one another. Such innovations democratize energy management, making it more inclusive and efficient.

In conclusion, the integration of AI in smart energy management offers a revolutionary approach to how we consume and distribute energy. Through predictive analytics, real-time monitoring, and dynamic adjustments, AI helps in making our energy systems more efficient, reliable, and sustainable. As technology continues to advance, the potential for AI-driven energy management will only grow, opening new avenues for innovation and sustainability. By embracing these intelligent systems, we are not only reducing our energy costs but also taking significant steps towards a sustainable future.

AI in Waste Management

Environmental sustainability is becoming increasingly critical as the world grapples with rising pollution levels, climate change, and resource depletion. One of the most promising areas of innovation in this sphere is AI's application in waste management. Modern technology offers tools that not only streamline waste processing but also make it smarter and more efficient. When implemented thoughtfully, AI can help us tackle one of the biggest challenges humanity faces: waste.

Waste management encompasses several stages, from waste generation to collection, transport, treatment, and disposal. Each stage presents unique challenges and opportunities for optimization. Traditionally, waste management systems have been manual and labor-intensive, leading to inefficiencies and higher costs. Enter artificial intelligence, which is poised to revolutionize this field by enabling smarter and more adaptable waste management systems.

One way AI enhances waste management is through intelligent sorting systems. Current recycling efforts are often hampered by contamination — improper sorting of recyclable materials. AI-based solutions can improve sorting accuracy significantly. For example, advanced machines equipped with computer vision and machine learning algorithms can identify different types of waste materials on a conveyor belt. These systems can discern between plastics, paper, metals, and organics with greater precision than manual sorting, thereby reducing contamination and improving recycling rates.

Another exciting application of AI is in predictive analytics. Waste collection schedules in cities are typically fixed, leading to situations where bins overflow or are collected while only partially full. AI can optimize route planning for waste collection vehicles by predicting waste generation patterns. Using data from sensors placed in trash bins, AI can determine the most efficient collection schedules and routes, reducing fuel consumption, labor costs, and carbon emissions.

AI also plays a significant role in waste reduction at the source. Businesses and households generate substantial amounts of waste, much of which could be minimized with smarter practices. AI-powered platforms can help consumers and companies track their waste footprint more accurately and offer actionable insights to reduce waste. For instance, restaurants can use AI to analyze customer order patterns and optimize inventory management, reducing food waste by ensuring ingredients are used before they spoil.

Treatment and disposal of waste also benefit from AI advancements. In waste-to-energy facilities, where waste is converted into energy through various processes, AI can help optimize operations. Machine learning algorithms can analyze data from various parameters within the facility to maximize energy output and minimize harmful emissions. AI-driven systems can also ensure that any hazardous waste is treated and disposed of in the safest manner possible.

In the broader context of environmental sustainability, AI in waste management supports the circular economy concept, promoting resource reuse, recycling, and reduction. By making waste management smarter and more efficient, AI helps maintain the balance between resource extraction and waste generation, contributing to a more sustainable future.

In particular, smart cities are increasingly adopting AI-driven waste management solutions to enhance urban living standards. Cities like Singapore and Copenhagen are leveraging AI to monitor and manage waste more effectively, using smart bins, predictive analytics, and automated collection systems. These initiatives demonstrate the potential of AI to transform urban waste management, leading to cleaner cities and a reduced environmental footprint.

AI's involvement in waste management isn't restricted to just urban environments, though. Remote and rural areas also benefit from these intelligent systems. For example, drones equipped with AI can monitor illegal dumping sites in remote regions, helping authorities identify and act against such practices promptly. Moreover, AI-driven waste management solutions can be scaled down and adapted to smaller communities, ensuring that waste processing is efficient even in less populated areas.

Another intriguing application is the use of AI to engage and educate the public about better waste management practices. Mobile apps utilizing AI can provide users with real-time guidance on how to properly sort and dispose of their waste, making community-wide participation more effective. Gamification elements in these apps can incentivize users to engage more actively in recycling and waste reduction efforts.

Moving forward, the challenge lies in integrating AI with existing waste management systems and overcoming the initial financial and technical barriers. Investment in AI infrastructure and training for personnel is necessary to ensure successful implementation. Collaboration between policymakers, businesses, and technology providers will be essential to harness AI's full potential in waste management.

Despite these challenges, the benefits far outweigh the obstacles. With AI, we can transition from reactive waste management to proactive waste optimization, contributing significantly to environmental sustainability. By addressing the inefficiencies in current systems, AI enables us to manage waste better, ensuring cleaner, greener communities for future generations.

In conclusion, AI offers incredible prospects for the future of waste management. From intelligent sorting and predictive analytics to enhancing public engagement, AI is set to redefine how we manage waste. As we continue to innovate and integrate AI into waste management, we pave the way for a more sustainable and resilient planet.

Environmental Monitoring

Environmental monitoring is increasingly becoming a high-tech endeavor, powered by advances in artificial intelligence. AI's ability to process and analyze vast amounts of data in real-time is proving instrumental in tackling environmental challenges. Robust environmental monitoring systems leverage AI to track air and water quality, biodiversity, and even predict natural disasters. This can have a transformative impact on how we manage and preserve natural ecosystems, making our world a more sustainable place to live.

One of the most compelling uses of AI in environmental monitoring is in the realm of air quality management. Urban areas around the globe suffer from severe air pollution, leading to health issues and reduced quality of life for their inhabitants. Traditional methods of air quality monitoring involve placing a limited number of sensors in specific locations, which often fail to provide a comprehensive picture. AI changes the game by integrating data from numerous sources, such as satellite imagery, traffic sensors, and social media feeds, to offer more accurate and widespread air quality assessments.

These AI-powered models can not only detect current pollution levels but also predict future trends. By leveraging machine learning algorithms, cities can preemptively manage air quality during critical times, such as high traffic periods or industrial activities. For example, advanced algorithms can identify pollution hotspots and recommend measures to mitigate their impact, such as controlling vehicle emissions or optimizing traffic flow.

Water quality monitoring is another area where AI is making significant strides. Contaminated water sources pose a serious threat to both public health and the environment. Traditional water testing methods can be slow and labor-intensive, often involving manual sampling and laboratory analysis. AI automates this process by using sensors and cameras to continuously monitor water bodies. Machine learning algorithms analyze this data in real-time, identifying pollutants and potential sources of contamination with high accuracy.

Beyond identifying current water quality issues, AI also offers predictive capabilities. For instance, AI models can predict algal blooms, which can devastate aquatic ecosystems and affect water supply. Such predictions can guide timely interventions to prevent ecological damage. Additionally, AI can optimize the operation of water treatment plants, ensuring that the treated water meets safety standards while minimizing resource consumption.

Biodiversity is another critical aspect of environmental monitoring enhanced by AI. Tracking animal populations and their habitats has traditionally been a labor-intensive task, often requiring human observation and manual logging. AI simplifies this by utilizing remote sensing technologies and computer vision algorithms. Drones equipped with cameras can survey large areas, capturing images that AI algorithms analyze to identify and count species. This allows for better management of wildlife reserves and more informed conservation strategies.

One fascinating application of AI in biodiversity is in acoustic monitoring. Many species, particularly birds and marine life, communicate through sounds that can be captured by microphones. AI algorithms are trained to recognize these sounds amidst background noise, offering valuable data on species presence and behavior. Such comprehensive monitoring is crucial for assessing the health of ecosystems and identifying areas that need conservation efforts.

Natural disaster prediction and management is another sphere where AI's capabilities are invaluable. Earthquakes, floods, and hurricanes pose severe risks to human life and the environment. AI systems analyze seismic data, weather patterns, and even historical disaster data to predict when and where natural disasters might occur. These predictions allow for timely evacuations and resource allocation, significantly reducing the potential harm.

In flood prediction, for instance, AI models integrate weather forecasts, topographical data, and river flow statistics to predict flooding events with remarkable precision. Authorities can use this information to issue early warnings and implement flood prevention measures. Similarly, in wildfire management, AI systems analyze satellite imagery and weather conditions to predict fire outbreaks, enabling quicker and more effective responses.

AI isn't just about predicting disasters; it's also pivotal in managing their aftermath. Post-disaster assessment often involves sifting through vast amounts of data to understand the impact and plan recovery efforts. AI accelerates this process by automating the analysis of satellite images, social media posts, and emergency response data, providing a clearer picture of the affected areas.

Moreover, AI's role in environmental monitoring isn't limited to large-scale initiatives. Community-level participation is also enhanced through AI-powered apps that allow citizens to contribute data. For instance, smartphone applications can use built-in sensors to measure air quality or photograph and report instances of illegal dumping. This citizen-generated data can be aggregated and analyzed to supplement official monitoring efforts, offering a more grassroots approach to environmental management.

One inspiring aspect of AI in environmental monitoring is its potential to drive policy decisions. Accurate and timely data puts pressure on policymakers to act on environmental issues more decisively. For instance, AI-driven insights into air and water quality can influence regulations on industrial emissions and waste management practices. This data-centric approach ensures that policies are based on real-world conditions, making them more effective and sustainable in the long run.

The integration of AI into environmental monitoring also opens up new avenues for international cooperation. Climate change and environmental degradation are global challenges that require collective action. AI systems can harmonize data from different countries, providing a unified perspective on global environmental health. This synchronized data can form the basis for international agreements and coordinated actions to address climate change, deforestation, and other pressing issues.

In conclusion, the potential of AI in environmental monitoring is vast and transformative. From improving air and water quality to tracking biodiversity and predicting natural disasters, AI offers innovative solutions to some of the most daunting environmental challenges. As AI technologies continue to advance, their application in environmental monitoring will undoubtedly become more sophisticated and integrated, offering us unprecedented opportunities to steward our planet more effectively. By leveraging AI's capabilities, we can move towards a future where environmental sustainability is not just an ideal but a tangible reality.

CHAPTER 17:
AI IN GOVERNMENT AND PUBLIC SERVICES

In the realm of government and public services, Artificial Intelligence stands as a transformative force, poised to redefine how societies operate and thrive. From smart cities leveraging data and automation to optimize urban living, to enhanced public safety through predictive policing and emergency response systems, AI's potential is monumental. Moreover, digital government services driven by AI can streamline bureaucratic processes, making them more efficient and transparent for citizens. This convergence of intelligent technology and public infrastructure not only promises to enhance service delivery but also aims to foster a more connected, safe, and accessible society for all.

Smart Cities

Imagine strolling through a city where the streetlights brighten as you approach, traffic aligns to your optimal route, and public services respond proactively to your needs. This isn't a sci-fi dream, but the burgeoning reality of smart cities, driven by advancements in artificial intelligence. By integrating AI into urban planning and management, cities are transforming into efficient, responsive environments that enhance the quality of life for their inhabitants.

At the core of a smart city lies its infrastructure. Traditional urban systems often operate in silos, but smart cities utilize a network of interconnected components, powered by AI, to create a cohesive, intelligent ecosystem. For instance, AI-driven sensors and IoT (Internet of Things) devices collect real-time data from various sources - traffic patterns, weather conditions, energy usage, and more. This vast reservoir of data allows city officials to make informed decisions, optimize resources, and anticipate problems before they escalate.

Traffic management is a quintessential application of AI in smart cities. Congestion is a common bane of urban life, leading to wasted time and increased pollution. Advanced traffic management systems harness AI to analyze traffic flow data and modify traffic signal timings dynamically, ensuring smoother commutes. Moreover, AI algorithms can predict traffic bottlenecks and suggest alternative routes to drivers via mobile apps, effectively reducing the overall traffic load.

Public transportation systems also benefit from AI integration. By analyzing usage patterns and real-time data, AI can optimize bus and train schedules, decrease wait times, and improve reliability. This not only enhances passenger experience but also boosts the efficiency of public transit, making it a more attractive option compared to private vehicles. Additionally, AI can manage predictive maintenance for transportation fleets, minimizing breakdowns and extending the lifespan of transit assets.

One cannot discuss smart cities without mentioning energy management. Urban areas are significant consumers of energy, and efficient management is crucial for sustainability. AI-driven smart grids are designed to balance supply and demand dynamically, using real-time data to adjust energy distribution. These grids can also incorporate renewable energy sources seamlessly, ensuring a stable and sustainable energy supply. AI algorithms can further optimize energy use in buildings, reducing wastage and lowering costs for both city authorities and residents.

Security is another critical domain where AI proves invaluable. Smart surveillance systems equipped with AI can monitor vast areas with minimal human intervention. By analyzing video feeds in real-time, these systems can detect unusual activities, identify potential security threats, and alert authorities promptly. Furthermore, AI-powered facial recognition technologies can enhance public safety by identifying known offenders or missing persons in crowded places.

Waste management, often overlooked, can also be revolutionized through AI. Smart bins equipped with sensors can signal when they need to be emptied, optimizing route planning for garbage collection trucks and reducing operational costs. AI can also analyze patterns in waste generation, enabling cities to implement more effective recycling and reduction strategies.

Public health, especially in dense urban areas, can significantly benefit from AI. Predictive analytics can forecast potential outbreaks of diseases, allowing for proactive measures to be taken. AI can also optimize the placement of healthcare facilities and resources based on projected demand, ensuring that healthcare services are both accessible and efficient. During times of crisis, such as pandemics, AI can play a pivotal role in managing the spread of diseases by offering timely data and predictions.

AI's impact on smart city architecture extends into environmental monitoring and sustainability efforts. By constantly monitoring air quality, water levels, and soil conditions, AI systems can provide critical insights that enable cities to respond swiftly to environmental concerns. These systems are also vital in disaster management, providing predictive analytics and real-time data to coordinate effective responses to natural calamities.

Consider the concept of a "digital twin" - a virtual replica of the entire city, updated in real-time using data from various sources. This digital twin, powered by AI, allows city planners and officials to simulate different scenarios, test potential solutions, and make data-driven decisions that enhance urban living. It's like having a city-wide crystal ball, offering unprecedented foresight and planning capabilities.

Communities benefit from smart city technologies in multifaceted ways. Smarter infrastructure leads to reduced operational costs and lower taxes, while enhanced services improve residents' quality of life. Moreover, the intelligent optimization of resources contributes to environmental sustainability, creating healthier urban environments. Public engagement platforms, powered by AI, can also facilitate better communication between city officials and residents, encouraging collaborative governance and increased civic participation.

However, the transition to smart cities does come with its challenges. Integration of AI into existing infrastructure can be costly and requires meticulous planning. Data privacy and security are paramount concerns, as the increased collection and analysis of personal data bring forth potential risks. Ethical considerations, such as ensuring equitable access to smart city benefits across different socioeconomic groups, also need to be addressed.

Nevertheless, many cities around the globe are successfully implementing AI-driven solutions, setting benchmarks for the future. Cities like Singapore, Barcelona, and Copenhagen are at the forefront of this revolution, showcasing how technology can redefine urban living. Their experiences offer valuable insights and best practices that other cities can emulate.

In conclusion, the advent of AI in smart cities represents a paradigm shift in urban living. By creating an intelligent, interconnected ecosystem, cities can achieve unparalleled efficiency, sustainability, and quality of life. While there are hurdles to overcome, the potential rewards far outweigh the challenges. As AI technologies continue to evolve, smart cities will not just be a trend but a cornerstone of future urban development, setting a new standard for how cities can—and should—operate.

AI in Public Safety

Artificial Intelligence (AI) is fundamentally transforming public safety, making communities not only safer but also more efficient and responsive. Imagine a world where emergency services can predict and prevent incidents before they occur, or where first responders have real-time information at their fingertips, enabling them to act swiftly and accurately. This is the promise of AI in public safety—a confluence of automation, data analytics, and machine learning designed to protect and serve more effectively than ever before.

One of the most profound impacts of AI in public safety is in predictive policing. By analyzing vast amounts of data, including crime reports, social media activity, and other public records, AI systems can identify patterns and predict where crimes are likely to occur. This allows law enforcement agencies to allocate resources more strategically, potentially preventing criminal activity before it starts. These AI algorithms can detect anomalies in crime data, help pinpoint hotspots, and even predict the likelihood of reoffending, enabling a proactive rather than reactive approach to law enforcement.

Enhanced surveillance is another area where AI is making significant strides. With AI-driven video analytics, surveillance cameras can now do much more than merely record footage. They can identify suspicious behaviors, recognize faces, and even detect unattended bags in crowded places. AI can analyze these video streams in real-time, alerting security personnel to potential threats and significantly reducing response times. These capabilities are particularly valuable in high-risk public spaces such as airports, train stations, and large public events.

AI is also revolutionizing emergency response services. Traditional emergency dispatch systems often struggle with slow response times and inaccurate information. AI-powered systems can sift through 911 calls, social media posts, and other data sources to provide a comprehensive, real-time picture of an emergency situation. This enables dispatchers to make more informed decisions and ensures that first responders have the information they need to act quickly and effectively. For example, natural language processing (NLP) algorithms can analyze emergency calls in real-time, extracting critical information about the nature and location of the incident.

Emergency medical services (EMS) are another beneficiary of AI technologies. AI-driven applications can predict the most likely times and locations for medical emergencies by analyzing historical data, weather conditions, and other variables. This allows EMS teams to be strategically positioned, reducing response times and potentially saving lives. Furthermore, AI can assist paramedics by providing real-time data about the patient's condition, suggesting potential diagnoses, and even guiding them through complex medical procedures.

In fire departments, AI is being used to predict the outbreak and spread of fires, analyze building layouts, and recommend the most effective strategies for firefighting. Predictive analytics can help identify areas at high risk of fire based on factors such as weather conditions, vegetation, and historical data. Drones equipped with AI can provide real-time aerial imagery of fire scenes, helping firefighters to quickly assess the situation and make informed decisions. Additionally, AI can optimize the deployment of fire trucks and personnel, ensuring that resources are used efficiently and effectively.

Through machine learning and data analytics, AI is helping to manage public health crises. For instance, AI models can predict the spread of infectious diseases based on patterns in data from hospital records, public health reports, and even social media activity. These predictive capabilities allow public health officials to implement preventative measures more effectively, such as targeted vaccination campaigns or travel restrictions. In the wake of the COVID-19 pandemic, AI has been instrumental in contact tracing, symptom tracking, and predicting outbreak hotspots.

Another critical application of AI in public safety is in the area of disaster response and recovery. AI systems can analyze a multitude of data points—including weather forecasts, geospatial data, and social media activity—to predict natural disasters like hurricanes, floods, or earthquakes. Early warning systems can then alert communities about impending dangers, giving people more time to evacuate or take other preventive measures. During the recovery phase, AI can help coordinate relief efforts, ensuring that resources are allocated where they are most needed.

AI's role in cybersecurity is also crucial to public safety. With the rise of cyber threats, protecting digital infrastructure has become as important as safeguarding physical spaces. AI-driven systems can monitor network traffic, detect anomalies, and identify potential security breaches in real-time. Machine learning algorithms can analyze patterns of behavior to distinguish between normal and suspicious activities, thereby enabling more effective prevention of cyber-attacks. This is particularly important for critical infrastructure such as power grids, transportation systems, and water supplies, where a cyber-attack could have devastating consequences.

On the frontline of combatting terrorism, AI is proving to be a formidable ally. By analyzing data from various sources, including social media, financial transactions, and travel records, AI can identify potential threats and flag individuals who may pose a risk. These capabilities have been instrumental in thwarting terrorist plots and identifying radicalization patterns. AI can also assist in crisis management during terrorist attacks, providing real-time information to law enforcement and emergency services.

Despite these advancements, the deployment of AI in public safety is not without its challenges. Concerns about privacy, data security, and ethical use of AI are paramount. For instance, the use of facial recognition technology has sparked debates about privacy and civil liberties. While AI can enhance public safety, it is crucial to ensure that its use does not infringe upon individual rights or lead to biased or discriminatory practices. Policymakers, technologists, and the public must work together to establish guidelines and regulations that balance the benefits of AI with the protection of civil liberties.

In conclusion, the integration of AI into public safety is revolutionizing how we protect and serve our communities. From predictive policing and enhanced surveillance to smarter emergency response and disaster management, AI is enabling a more proactive, efficient, and informed approach to public safety. As these technologies continue to evolve, they hold the promise of creating safer communities, better-prepared emergency services, and a more resilient society. The future of public safety lies in harnessing the power of AI to anticipate, prevent, and respond to the challenges of an increasingly complex world.

Digital Government Services

The integration of AI in digital government services isn't just a trend; it's a revolution poised to make governance more efficient, transparent, and accessible. Let's consider the ubiquitous tasks of filing taxes, renewing a driver's license, or even voting. AI is streamlining these services, transforming them into user-friendly experiences that can be accessed anytime, anywhere.

One of the most prominent applications of AI in digital government services is through virtual assistants. These AI-driven tools are employed to answer citizen inquiries, guide them through complex processes, and provide real-time support. Think of it as an administrative assistant available 24/7, ready to resolve your queries and ensure that you complete your tasks without unnecessary hassles.

Imagine logging into a government portal and having an intelligent bot greet you by name. This bot isn't just for show; it's analyzing your past interactions to offer personalized assistance. Need to renew your passport? The bot can guide you through the required steps, fill out forms based on previously stored data, and even schedule an appointment for you—all while keeping your data private and secure.

Moreover, AI is enhancing the efficiency of internal government operations. Machine learning algorithms can analyze vast amounts of data to predict trends, highlight areas requiring intervention, and optimize resource allocation. For instance, by monitoring social media and public forums, AI can gauge public sentiment about policies and services, offering invaluable insights to policymakers.

In the realm of public health, AI's ability to analyze complex datasets can be harnessed for everything from tracking the spread of infectious diseases to managing vaccination programs. During the COVID-19 pandemic, various governments deployed AI algorithms to model virus spread, inform lockdown measures, and even to assist in vaccine distribution logistics.

Additionally, AI is bringing about a paradigm shift in the realm of public safety. Using AI-driven surveillance systems, governments can ensure safer public spaces by detecting unusual activities and alerting law enforcement in real time. These systems can differentiate between false alarms and genuine threats, enhancing responsiveness while minimizing human error.

AI-driven fraud detection systems are particularly revolutionary in social welfare programs. These systems analyze patterns to identify inconsistencies and irregularities, ensuring that resources are directed to those who genuinely need them. This reinforces the integrity and sustainability of such programs, bolstering public trust in government institutions.

The promise of AI doesn't stop at domestic governance; it's also improving international relations. Through AI, governments can quickly translate documents and communications between different languages, facilitating more effective diplomacy and international cooperation. AI algorithms can also predict geopolitical trends, thereby contributing to more strategic foreign policy decisions.

Another intriguing application is in e-governance transparency. Blockchain technologies, combined with AI, are being used to create immutable records of government transactions, reducing corruption and improving accountability. For example, each step of a public procurement process can be logged and made transparent, ensuring that every cent is traceable.

The benefits also extend to legislative processes. AI can analyze past legislative records, debates, and outcomes to predict the impact of proposed laws. This allows lawmakers to make data-driven decisions, potentially increasing the effectiveness and fairness of new regulations.

Of course, these advancements come with challenges—data privacy being the foremost concern. As governments collect more data to fuel AI models, they must prioritize robust data protection mechanisms to safeguard citizens' personal information. Striking a balance between leveraging data for improvement and respecting individual privacy is crucial for the sustained success of digital government services.

In terms of public engagement, the use of chatbots and intelligent interfaces makes navigating governmental websites straightforward, even for those less technologically adept. Accessibility features, powered by AI, ensure that these services are inclusive, accommodating various languages, and assisting citizens with disabilities through voice commands and other adaptive technologies.

Moreover, AI-powered predictive analytics plays a vital role in disaster management. By analyzing historical data along with real-time feeds from satellites and sensors, AI can help governments predict natural disasters such as floods, earthquakes, or hurricanes. This enables preemptive actions such as timely evacuations and resource allocations, potentially saving countless lives.

The educational sector, too, benefits from AI in government services. AI can identify regions where educational resources are lacking and allocate funding and support where it's needed most. It can also assist in developing tailored educational programs that cater to the unique needs of different communities.

Future prospects for digital government services look promising, with AI advancing at an exponential rate. The integration of emerging technologies such as 5G and the Internet of Things (IoT) will further enhance the capabilities and reach of AI in governance. We can anticipate a future where government services are so seamlessly integrated into our daily lives that they become almost invisible, operating efficiently in the background.

Collaboration between private companies and governments is essential in driving these innovations. Tech firms specializing in AI can offer their expertise to governmental bodies, ensuring that the most advanced and effective solutions are deployed. This public-private synergy can create robust, scalable solutions addressing the multifaceted challenges of modern governance.

Ultimately, the adoption of AI in digital government services isn't just a step towards modernization; it's a leap towards creating a more equitable, efficient, and responsive governance structure. This transformation has the potential to redefine our interaction with public services, making them more intuitive, less burdensome, and incredibly effective.

As we embrace these advancements, it's imperative to continuously evaluate their impact, iterate on feedback, and ensure inclusivity. Only then can we realize the full potential of AI in crafting a more connected, informed, and empowered society.

CHAPTER 18:
AI IN LEGAL SERVICES

The integration of AI in legal services is transforming a traditionally labor-intensive field, making it more efficient and accessible. AI-driven legal research tools can quickly analyze vast databases of case law, statutes, and legal documents, providing lawyers with comprehensive insights at unprecedented speeds. Intelligent contract analysis automates the review of lengthy legal agreements, identifying key clauses and potential issues, thus reducing the time and costs associated with manual review. Moreover, predictive analytics can forecast case outcomes based on historical data, aiding attorneys in crafting more effective strategies and making informed decisions. By streamlining these processes, AI not only increases productivity but also enhances the accuracy and consistency of legal work, ultimately benefiting clients and practitioners alike.

AI in Legal Research

AI in legal research has become a transformative force in the legal industry. Traditionally, legal research has been a time-consuming process requiring sifting through vast amounts of information to find relevant cases, statutes, and regulations. AI simplifies this by leveraging advanced algorithms to process and analyze data at an unprecedented speed, significantly reducing the time and effort needed to conduct thorough research.

One of the primary advantages of AI in legal research is its ability to process large volumes of text rapidly. AI systems use Natural Language Processing (NLP) to understand and interpret legal language, enabling them to scan through hundreds of documents in seconds. This capability not only expedites the research process but also ensures a higher degree of accuracy, as AI can identify pertinent information that may be overlooked by human researchers.

The deployment of AI in legal research is not just about speed and efficiency. It also introduces a higher level of precision. AI tools can identify and analyze complex legal principles, contradictions, and similarities across various jurisdictions. This is particularly beneficial in cases involving numerous laws and precedents, where human researchers might miss subtle but critical nuances.

Another notable aspect of AI in legal research is the ability to provide predictive insights. By examining historical legal data, AI can predict outcomes based on previous case law, helping lawyers build more effective strategies. For instance, knowing how similar cases have been decided can guide legal professionals in formulating arguments and advising their clients. This predictive capability can significantly affect the planning and management of legal cases.

AI tools offer enhanced search functionalities, including semantic search, which is more intuitive than traditional keyword-based search. With semantic search, AI understands the context and meaning behind search queries, delivering results that are more relevant and comprehensive. This leads to more effective and efficient legal research, allowing lawyers to find the most pertinent information without wading through unrelated data.

Legal research also benefits from AI's capacity for continuous learning. As AI systems process more data, they become more adept at recognizing patterns and drawing connections. This makes them increasingly effective over time, continually improving the quality and relevance of the information they provide. The self-improving nature of AI ensures that legal research remains at the cutting edge, keeping pace with the ever-evolving legal landscape.

Collaboration platforms powered by AI are another exciting development in legal research. These tools enable legal professionals to work together seamlessly, sharing insights and findings in real-time. AI can facilitate collaboration by highlighting relevant information, suggesting research avenues, and even summarizing complex legal documents. This fosters a more integrated and efficient research environment, enhancing the overall productivity of legal teams.

The ethical implications of AI in legal research cannot be overlooked. While AI can greatly improve efficiency and accuracy, it is crucial to ensure that these systems are unbiased and transparent. Developers must create AI tools that adhere to strict ethical guidelines, minimizing biases and ensuring that the technology supports fair and just legal practices. Safeguarding against biases in AI algorithms is especially vital in the legal field, where impartiality is paramount.

Moreover, the integration of AI in legal research paves the way for a more equitable legal system. By making legal research more accessible and affordable, AI democratizes access to legal knowledge. Smaller law firms and individual practitioners, who may not have extensive resources, can leverage AI to compete on a more level playing field with larger institutions. This can lead to broader access to legal services for individuals and communities that have traditionally been underserved.

The future of AI in legal research looks promising. As technologies like machine learning and NLP advance, AI's capabilities in legal research will continue to grow. Future developments might include more sophisticated predictive analytics, which could anticipate legal trends and shifts in judicial thinking. Additionally, AI could play a crucial role in international law by cross-referencing and analyzing legal systems across different countries, aiding in global legal practices.

AI in legal research is revolutionizing the legal field by making research faster, more accurate, and more accessible. It allows legal professionals to focus on higher-level tasks, such as strategy and client consultation, rather than spending extensive hours on manual research. The potential for AI to transform legal research is immense, promising a future where legal professionals are empowered by advanced technology to deliver better outcomes for their clients and the justice system as a whole.

In summation, the advent of AI in legal research signifies a pivotal shift in how legal services are delivered. By embracing these technologies, legal professionals can enhance their practice, provide more informed advice, and contribute to a just and efficient legal system. As AI continues to evolve, its role in legal research will only become more integral, shaping the future of the legal industry in profound and positive ways.

Intelligent Contract Analysis

Artificial Intelligence has transformed many sectors, but its impact on legal services is particularly significant. With the advent of Intelligent Contract Analysis, legal professionals now have tools that can streamline the traditionally complex and labor-intensive process of contract review and management.

What is Intelligent Contract Analysis, you may wonder? At its core, it involves the use of AI algorithms to evaluate and interpret legal agreements. This technology is designed to recognize key terms, obligations, deadlines, and potential risks, allowing lawyers to focus on strategic decision-making rather than getting bogged down in administrative tasks. Imagine not having to spend hours combing through dense legal jargon; instead, an AI can highlight the critical points and even flag potential issues.

One of the most impactful benefits of Intelligent Contract Analysis is its ability to increase accuracy. Human error is an inevitable part of manual contract review, especially when dealing with large volumes of documents. No one is immune to fatigue or occasional oversight. AI systems, however, don't tire. They can work around the clock, and their precision doesn't waver with time. This is particularly useful in due diligence processes, where meticulous review is non-negotiable.

These AI systems are not just about parsing through the text and identifying clauses; they also compare contracts against applicable laws and regulations. This ensures compliance, reducing the risk of legal disputes and fines. Moreover, by keeping abreast of the latest legal updates, Intelligent Contract Analysis tools can adapt to new legal requirements almost instantly, a task that would be cumbersome for human experts to manage manually.

Furthermore, Intelligent Contract Analysis can be a game-changer in negotiations. Consider the scenario where two parties are drafting a contract: AI can help identify mutually agreeable terms by analyzing preferences and past agreements. This capability facilitates smoother negotiations and faster resolutions. Automation doesn't just offer efficiency; it brings about better outcomes through data-driven insights.

Another transformative aspect lies in contract lifecycle management. Once a contract is signed, it's crucial to track obligations and deadlines to ensure ongoing compliance. AI can manage these tasks seamlessly, sending alerts and reminders for key dates and obligations. Imagine having an intelligent assistant that keeps tabs on every commitment, deadline, and renewal date for hundreds or thousands of contracts at once. This proactive management helps prevent breaches and fosters stronger business relationships.

Legal teams often face pressure to remain cost-effective while handling increasing workloads. AI in legal services, particularly through Intelligent Contract Analysis, frames a compelling case for reducing operational costs. By automating routine tasks associated with contract review and compliance, firms can allocate resources more efficiently, allowing their staff to concentrate on higher-value activities. The cost-savings can be dramatic, essentially by cutting down on the man-hours required for repetitive tasks.

Next, consider the scalability of AI-driven solutions. Whether you're a small firm with dozens of contracts or a large corporation managing thousands, Intelligent Contract Analysis scales effortlessly. Such tools can process large volumes in a fraction of the time it would take human review teams. This capability is invaluable in mergers and acquisitions, corporate reorganizations, and other scenarios involving substantial documentation.

Another critical advantage is the potential for enhanced collaboration. Modern AI-driven tools often integrate easily with other software platforms, enabling seamless sharing and real-time collaboration. For instance, imagine a cloud-based contract management system that allows different stakeholders, from legal to procurement to sales, to access and work on the same document simultaneously. This level of transparency and collaboration can lead to more cohesive and synchronized operations.

Now, let's talk about the sophistication of AI algorithms involved in Intelligent Contract Analysis. Natural Language Processing (NLP) techniques, combined with machine learning, are at the heart of these systems. NLP enables the AI to understand and interpret the language used in contracts, while machine learning improves the system's accuracy over time. As more contracts are analyzed, the algorithms learn from that data, continually refining their understanding. Essentially, the more you use it, the better it becomes.

Of course, the use of Intelligent Contract Analysis is not without challenges. One significant concern is data security. Legal documents often contain highly sensitive information, and ensuring the confidentiality of this data is paramount. Advanced encryption methods and secure access protocols must be in place to mitigate these risks. Furthermore, there needs to be a balance between leveraging AI for efficiency and maintaining the essential human oversight required for nuanced legal interpretation.

Another challenge pertains to the interpretability of AI decisions. Legal professionals might be skeptical of relying on an AI system if they don't understand how it arrived at a particular conclusion. Addressing this requires transparency in how AI algorithms function, often referred to as "explainable AI." It's not just about the what but also the why, ensuring that the technology can be trusted and its recommendations acted upon confidently.

Despite these challenges, the advantages of Intelligent Contract Analysis far outweigh the drawbacks. The technology is still evolving, continually improving in accuracy, reliability, and security. By embracing these advancements, legal professionals can navigate the complexities of their field with greater ease and effectiveness. The future-ready law firm isn't just about having skilled attorneys; it's about integrating intelligent systems that enhance decision-making and operational efficiency.

One shouldn't overlook the broader impact of Intelligent Contract Analysis on the legal profession as a whole. As these advanced technologies become standard, the skills required in the legal field may evolve. Future legal professionals might need to be proficient not just in law but also in understanding and leveraging AI tools. This shift opens up new educational and career opportunities, encouraging a generation of tech-savvy lawyers who can marry legal expertise with technological acumen.

So, the next time you think of the often daunting task of contract analysis, remember that AI is here to assist. It's transforming how contracts are reviewed, analyzed, and managed, making life easier for legal professionals and benefiting clients through more precise, efficient, and reliable services. The legal industry is not merely adapting to the AI revolution; it is proactively embracing and shaping it, setting the stage for a future where technology and human expertise coalesce into a powerful force.

Predictive Analytics in Law

Predictive analytics is transforming a variety of fields, and the legal sector is no exception. With the infusion of artificial intelligence, legal services can now benefit from unprecedented levels of efficiency and foresight. Predictive analytics involves using historical data and complex algorithms to predict future outcomes. In the context of law, this has profound implications for how cases are approached, how legal strategies are developed, and even how outcomes are forecasted.

Imagine being able to predict the likelihood of winning a court case before it even goes to trial. This is one of the powerful promises of predictive analytics in law. Utilizing vast databases of prior cases, legal precedents, and judge-specific tendencies, AI algorithms can approach each new case with a level of objectivity that would be nearly impossible for a human.

Take, for example, the process of litigation. Traditionally, lawyers have relied heavily on their experience and intuition to decide whether to take a case, settle, or go to trial. With predictive analytics, lawyers can now back their instincts with data-driven insights. This not only reduces the risk associated with litigation but also optimizes resource allocation. Knowing the statistical probability of various outcomes allows legal teams to channel their efforts more effectively and strategically.

Another crucial application of predictive analytics in law is in the realm of contract management. Contracts are the lifeblood of business transactions, and their complexity can sometimes be overwhelming.

AI-driven predictive analytics can assist in identifying potential red flags, assessing the risk of non-compliance, and even predicting the long-term performance of contractual relationships. By analyzing historical contract data, AI systems can provide actionable insights into which contracts are likely to face disputes and recommend preemptive steps to mitigate those risks.

The judicial system also stands to gain tremendously from predictive analytics. Judges and court administrators can use AI to manage caseloads more efficiently. For instance, predictive models can forecast the potential duration of a case, helping with docket management and reducing backlogs. Additionally, these predictive insights can aid in making data-backed decisions about parole, probation, and sentencing, ensuring a more balanced and fair judicial process.

One might argue that such reliance on predictive analytics might undermine the human element in law—which is undeniably profound and nuanced. This concern is valid, but it's essential to view AI as a complementary tool rather than a replacement. Predictive analytics extends the capabilities of legal professionals rather than curbing their judgment. By offering a data-driven foundation, AI helps ensure that human expertise is applied where it's most impactful.

Client relationship management is another area where predictive analytics is making waves. Understanding client behavior and predicting their needs can significantly enhance legal service delivery. AI can analyze patterns in client interactions, helping law firms anticipate inquiries, predict case outcomes, and tailor their communication strategies accordingly. This leads to a more personalized client experience, fostering stronger relationships and enhancing client satisfaction.

Consider the massive advantage law firms can gain in competitive intelligence. Through the power of predictive analytics, firms can examine trends and patterns in their competitors' behaviors. For instance, they can glean insights on how rival firms handle similar cases, what strategies they employ, and how successful they are. This data not only helps in benchmarking but also offers actionable intelligence to refine one's own strategies.

Moreover, the capabilities of predictive analytics extend to regulatory compliance. In industries that are heavily regulated, ensuring conformity to laws and guidelines is paramount. Predictive analytics can flag potential areas of non-compliance by analyzing existing regulations, past violations, and organizational behaviors. Legal departments can then proactively address these areas, minimizing the risk of legal repercussions and safeguarding the organization's integrity.

Legal research, traditionally a time-consuming task, is another domain where predictive analytics proves invaluable. By sifting through enormous quantities of legal texts, case laws, and journal articles, AI systems can identify relevant information more swiftly and accurately than humans. This not only accelerates the research process but also ensures comprehensiveness, leaving no stone unturned. Consequently, lawyers can dedicate more time to crafting robust legal arguments, armed with data-backed insights.

Additionally, it's worth noting that predictive analytics can also play a pivotal role in legal education. Aspiring lawyers can gain access to predictive models that help them understand the dynamics of real-world legal scenarios. By interacting with these models, they can simulate different case outcomes based on varying legal strategies. This provides a practical learning environment, bridging the gap between theoretical knowledge and practical application.

Ethical considerations surrounding the use of predictive analytics in law cannot be overlooked. Bias and fairness are significant concerns, given that AI systems learn from historical data that may contain inherent biases. Hence, it's crucial to ensure that these systems are designed and trained responsibly. Transparent algorithms, regular audits, and strict adherence to ethical guidelines can help mitigate these risks, ensuring that the use of predictive analytics promotes justice rather than undermining it.

Despite its immense potential, the adoption of predictive analytics in law isn't without challenges. Technological barriers, data privacy concerns, and the need for specialized skills are some of the hurdles that need to be overcome. However, the relentless advancement in AI technology and its increasing integration into various sectors provide a promising outlook for its application in legal services.

In conclusion, predictive analytics is revolutionizing the legal landscape by enabling data-driven decision-making, enhancing efficiency, and providing deeper insights. As this technology continues to evolve, its applications in law are likely to become even more sophisticated, offering legal professionals powerful tools to navigate complex legal challenges. By embracing predictive analytics, the legal sector can not only improve its operational efficiency but also contribute to a more transparent, fair, and just legal system.

CHAPTER 19:
AI IN MARKETING

In today's rapidly evolving digital landscape, AI has fundamentally transformed marketing by providing unprecedented abilities to analyze vast amounts of data and predict consumer behavior. Imagine harnessing the power of targeted advertising that understands your preferences better than you do, delivering personalized content and offers that are almost impossible to ignore. But it doesn't stop there—AI-driven customer insights delve deeper, uncovering patterns and trends that would take humans years to decipher. Content generation, another marvel, allows marketers to craft engaging and relevant materials at scale, ensuring messages resonate effectively with diverse audiences. Leveraging AI in marketing isn't just about enhancing efficiency; it's about building smarter, more intuitive, and impactful brand experiences that drive results and foster customer loyalty.

Targeted Advertising

The rise of Artificial Intelligence in marketing has revolutionized the way businesses reach and engage their customers. Perhaps one of the most transformative applications is in targeted advertising. This sophisticated form of advertising leverages AI to deliver personalized ads to users, maximizing engagement and conversion rates. By analyzing mountains of data, AI can understand the intricacies of human behavior, preferences, and needs far better than traditional methods. The result? Ads that feel more like recommendations from a friend than a hard sell from a corporation.

AI achieves this through a myriad of technologies, including machine learning, natural language processing, and predictive analytics. These technologies work harmoniously to sift through vast datasets, identifying patterns and trends that would be nearly impossible for humans to discern. This allows advertisers to craft messages that resonate on an individual level, delivering content that is relevant and timely. Imagine browsing a website and seeing an ad for a product you've been thinking about purchasing and then receiving a discount offer in your email moments later. Such experiences are not coincidental but are meticulously designed by AI algorithms.

One of the most direct benefits of AI in targeted advertising is its ability to create personalized experiences for consumers. Companies can now tailor their advertising strategies based on the specific preferences and behaviors of individual users. This level of personalization involves analyzing previous interactions, purchase history, and even social media activity to craft unique ads. For instance, an AI can identify a user who frequently shops for high-end fashion and then deliver ads showcasing the latest designer collections. The more precise the targeting, the higher the likelihood of engagement and conversion.

This capability extends beyond just direct interactions. AI can incorporate contextual information to further refine its targeting algorithms. Factors such as geographic location, time of day, and even the weather can influence consumer behavior, and modern AI systems can adjust ads based on these variables. A customer in New York might see ads for winter clothing during a cold spell, while someone in Miami might be targeted with beachwear promotions.

The implications for companies are profound. Targeted advertising not only improves ROI but also enhances customer satisfaction by reducing the noise of irrelevant ads. For consumers, this means fewer distractions and more meaningful interactions with brands. It's a win-win scenario that exemplifies how AI can create mutual benefits for businesses and their audiences.

Another critical component of AI in targeted advertising is real-time optimization. Traditional advertising campaigns often require significant lead time to plan and execute. However, AI can monitor campaign performance in real-time, making immediate adjustments to improve outcomes. If an ad isn't performing well, the AI can tweak elements such as the headline, imagery, or target audience instantaneously. This agility allows businesses to maximize their advertising spend and achieve better results faster than ever before.

To fuel these advanced targeting capabilities, AI engines must be fed a continuous stream of data. This brings us to another vital aspect: data collection and analysis. Companies today utilize AI to collect data from numerous sources, including websites, social media platforms, and customer databases. The more data an AI system has access to, the better it becomes at predicting consumer behavior and preferences. However, this also raises important considerations regarding data privacy and user consent, which we'll discuss in a later chapter.

With the increasing sophistication of AI, the ethical landscape of targeted advertising is also evolving. There's a thin line between providing personalized experiences and invading privacy. Transparency is key. Consumers should be made aware of how their data is being used and should have the ability to opt-out if they choose. Ethical AI ensures that while ads are targeted and personalized, they respect user boundaries and consent. Brands that prioritize ethical practices in their advertising strategies are likely to build more trust and loyalty among consumers.

In addition to individual targeting, AI also enables hyper-segmentation. Traditional marketing segmentation might categorize consumers based on broad demographic factors. In contrast, AI allows for the creation of highly specific micro-segments based on nuanced behavioral data. These micro-segments can be as detailed as users who buy running shoes every six months or those who only shop during holiday sales. By understanding these micro-segments, companies can create more targeted and effective ad campaigns, reaching the right people at the right time with the right message.

The impact of AI on targeted advertising is not limited to online platforms. Offline experiences, too, are being enhanced through AI. For instance, retail stores can use AI to analyze in-store behavior, such as how customers navigate the aisles or which products they pick up and put back. This data can then be used to optimize store layouts, personalize in-store promotions, and even subtly influence purchasing decisions. By bridging the gap between online and offline data, AI creates a cohesive and personalized shopping experience across all touchpoints.

Another exciting development is the integration of AI with augmented and virtual reality technologies. Imagine walking down a street and seeing personalized ads on digital billboards tailoring their message just for you. Or, while using a VR headset, interacting with virtual products that AI has determined fit your preferences. These immersive advertising experiences, powered by AI, add an entirely new dimension to how brands can engage with consumers.

AI's influence in targeted advertising is continually growing, driven by advancements in technology and increased data availability. As it becomes more integrated into advertising strategies, businesses that leverage AI will likely find themselves at a significant competitive advantage. The key is to stay ahead of the curve by adopting these technologies early, continually optimizing strategies based on AI insights, and maintaining a strong ethical stance on data usage.

The future of AI in targeted advertising looks promising, with potential innovations like predictive marketing, where AI anticipates consumer needs before they even realize them. This could result in more proactive advertising strategies that provide value and solutions precisely when needed. As AI evolves, so too will the methods and effectiveness of targeted advertising, ensuring this field remains an essential component of modern marketing.

In conclusion, targeted advertising powered by AI is transforming marketing into a more precise, personalized, and effective practice. By harnessing the power of data and machine learning, companies can deliver ads that truly resonate with their audience, creating meaningful and engaging experiences. As we move forward, the synergy between AI and targeted advertising will only become stronger, setting the standard for how businesses connect with their customers in the digital age.

Customer Insights

In the dynamic landscape of marketing, customer insights have always been the bedrock upon which successful strategies are built. Understanding what customers want, how they behave, and why they make certain choices can profoundly impact a company's ability to engage and retain its audience. This is where AI steps in, revolutionizing the way businesses gather, analyze, and leverage customer data.

AI has transformed the traditional methods of obtaining customer insights by automating the data collection process and enhancing the analysis capabilities. Machine learning algorithms can sift through vast amounts of data at speeds unimaginable a few years ago, identifying patterns and trends that human analysts might miss. For instance, AI can process social media interactions, online reviews, purchasing behaviors, and even customer service inquiries to form a comprehensive understanding of a customer's preferences and needs.

One of the most significant advantages of using AI for customer insights is the ability to achieve real-time analysis. Traditional methods often involve periodic surveys or focus groups, providing snapshots of customer sentiment at specific points in time. However, AI-powered tools can monitor customer interactions as they happen, allowing businesses to respond instantaneously to changing trends and preferences. This agility can be a game-changer in highly competitive markets.

Furthermore, AI enables more accurate segmentation of customers. By analyzing various data points—such as demographic information, purchasing history, and online behavior—AI can identify subgroups within a larger audience. These segments can then be targeted with personalized marketing strategies, enhancing the relevance and effectiveness of marketing campaigns. Personalization, driven by AI insights, can lead to increased customer satisfaction and loyalty.

AI-driven customer insights also play a crucial role in predictive analytics. Predictive models can forecast future behaviors based on historical data, providing businesses with a glimpse into what their customers might do next. For example, if an e-commerce platform knows that a particular customer tends to buy electronics during the holiday season, it can proactively send personalized offers on related products. This level of foresight can significantly boost sales and reduce churn.

The natural language processing (NLP) capabilities of AI also contribute to understanding customer sentiment at a granular level. NLP algorithms can analyze text data from various sources—such as reviews, social media posts, and customer service transcripts—to gauge the sentiment behind the words. This sentiment analysis allows businesses to measure the emotional tone of customer interactions, providing valuable insights into customer satisfaction and areas needing improvement.

An often overlooked but critical aspect of AI in customer insights is the ethical use of data. Companies must navigate the fine line between leveraging customer data for insights and respecting privacy.

Transparent data practices and adherence to privacy regulations are essential in building and maintaining customer trust. AI technologies can assist in this by anonymizing data and implementing robust security measures, ensuring that customer information is used responsibly.

The integration of AI in customer insights is not limited to large corporations. Small and medium-sized enterprises (SMEs) can also benefit from AI-powered tools designed to fit different scales of operations and budgets. Many AI solutions offer user-friendly interfaces and intuitive dashboards, making it accessible for businesses of all sizes to harness the power of AI without extensive technical expertise.

Moreover, the continuous learning aspect of AI means that these systems become more accurate and valuable over time. As more data is collected, machine learning models refine their algorithms, leading to even more precise and actionable insights. This ongoing improvement can help businesses stay ahead of the curve in understanding evolving customer behaviors and market trends.

One real-world example of AI-driven customer insights is the use of smart recommendation engines in e-commerce. These engines analyze a customer's past purchases and browsing history to recommend products they might be interested in. Companies like Amazon have perfected this technique, which not only boosts sales but also enhances the customer experience by making shopping more convenient and personalized.

Another example can be seen in the finance sector, where AI analyzes customer spending patterns to offer tailored financial advice. By understanding a customer's income, expenses, and saving habits, AI can suggest personalized investment plans or budgeting tips. Such insights can empower customers to make better financial decisions and improve their financial well-being.

In conclusion, AI's role in customer insights marks a significant shift in how businesses approach marketing. By providing real-time, accurate, and actionable data, AI enables companies to develop more effective marketing strategies, offer personalized experiences, and anticipate customer needs. As technology continues to evolve, the ways in which businesses can harness AI for customer insights will only expand, opening new horizons for engagement and growth.

The integration of AI in customer insights is not just a technological advancement; it represents a paradigm shift in understanding and meeting customer needs. By leveraging the power of AI, businesses can create more meaningful connections with their customers, fostering loyalty and driving long-term success.

Content Generation

In the rapidly evolving world of AI, content generation stands out as a transformative application in marketing. At its core, AI-driven content generation leverages algorithms and natural language processing (NLP) to create text, images, and even videos, reshaping traditional marketing strategies. This capacity to generate personalized, high-quality content quickly and efficiently is revolutionizing how brands communicate with their audiences.

The true power of AI in content generation lies in its ability to analyze vast amounts of data and discern patterns that human marketers might miss. When creating content, AI tools consider factors such as customer preferences, engagement metrics, and even cultural trends, ensuring that the output is not only relevant but also compelling. For businesses, this means more targeted and effective marketing campaigns, increasing engagement and conversion rates.

One of the most prominent applications of AI in content generation is in writing and editing. Tools like OpenAI's GPT-3 have shown remarkable prowess in producing written content that is indistinguishable from human-authored text. Whether it is blog posts, social media updates, or email newsletters, AI can draft content that meets specific brand guidelines and tone-of-voice requirements. Marketers can then refine the generated drafts, significantly reducing the time and effort required for content creation.

Moreover, AI-powered platforms such as Copy.ai and Writesonic offer functionalities that go beyond mere text generation. These tools can craft catchy headlines, write engaging product descriptions, and generate full-length articles that align with SEO best practices. By automating these tasks, businesses can ensure a consistent flow of high-quality content while freeing up human resources for more strategic functions.

Visual content is another area where AI is making substantial inroads. Tools like Artisto and Runway ML enable the creation of visually stunning graphics and videos with minimal human intervention. By analyzing image data, these AI applications can generate visuals that capture the essence of a brand's message. For example, AI can design social media posts or promotional videos that are not only visually appealing but also tailored to the target audience's preferences. This capability is particularly beneficial for smaller businesses and startups that may not have the budget for a full-fledged design team.

Beyond text and visuals, AI is also transforming how interactive content is created. Chatbots powered by AI can simulate human-like conversations, providing a personalized experience for the user. These chatbots can be integrated into websites and social media platforms, guiding users through product selections, answering queries, and even processing orders. By handling these interactions, AI-driven chatbots enhance customer engagement while reducing the workload on human customer service representatives.

Another fascinating aspect of AI in content generation is its role in content curation. AI algorithms can sift through vast amounts of data, including user-generated content, to identify the most relevant and engaging pieces. This curated content can then be shared with the audience, providing a personalized experience that drives engagement. Platforms like Curata and ContentStudio utilize AI to discover, organize, and distribute content that resonates with specific audience segments, making the content marketing process more efficient and effective.

AI's ability to generate multilingual content is also a game-changer for global marketing campaigns. Solutions like Google Translate have evolved significantly, providing more accurate translations that retain the context and nuances of the original text. By integrating AI-driven translation tools, businesses can produce content that caters to diverse linguistic audiences without compromising on quality. This capability not only broadens the reach of marketing efforts but also enables brands to connect with audiences on a more personal level.

In the realm of email marketing, AI-driven content generation tools are transforming how campaigns are designed and executed. AI can analyze recipient data to craft personalized email content that addresses individual preferences and behaviors. Predictive analytics further enhance this process by forecasting the best times to send emails and the types of content that are likely to engage each recipient. By automating these tasks, AI ensures that email marketing campaigns are not only more effective but also more efficient, maximizing ROI.

Furthermore, AI-generated content can be tested and optimized in real-time. A/B testing, long heralded in the marketing world, has been supercharged by AI's analytical capabilities. Algorithms can quickly analyze the performance metrics of different content variations and automatically iterate to improve outcomes. Platforms like HubSpot and Marketo incorporate AI to facilitate continuous testing and optimization, ensuring that the most effective content is always presented to the audience.

Voice is another frontier where AI is making waves in content generation. With the rise of smart speakers and voice assistants, creating audio content that is both engaging and informative has become increasingly important. AI-driven platforms like Replica and Lyrebird can generate human-like voiceovers for various marketing materials, including ads, tutorials, and podcasts. These tools allow brands to create personalized and interactive audio experiences that engage audiences in a novel and impactful way.

The seamless integration of AI in content generation also brings to light ethical considerations. While AI can produce vast amounts of content quickly, it is crucial to ensure that this content adheres to ethical guidelines and avoids misinformation. AI tools must be programmed to respect copyright laws and intellectual property rights, and businesses must ensure that AI-generated content is transparent about its origins. Ethical AI use fosters trust and authenticity in marketing efforts, where maintaining integrity is paramount.

Looking ahead, the potential for AI in content generation is boundless. Emerging technologies like deep learning and generative adversarial networks (GANs) promise even more sophisticated content creation capabilities. These advancements could enable AI to produce hyper-realistic videos and 3D graphics, opening new possibilities for immersive marketing experiences. As AI continues to evolve, marketers must stay abreast of these technological developments to leverage them effectively in their strategies.

In essence, AI-driven content generation is revolutionizing the marketing landscape by enabling brands to create personalized, engaging, and high-quality content with unprecedented efficiency. By automating and enhancing various aspects of the content creation process, AI empowers marketers to focus on strategic initiatives that drive growth and innovation. As you navigate the future of marketing, embracing AI in content generation can lead to more impactful and meaningful connections with your audience, driving success in an increasingly digital world.

CHAPTER 20:
AI IN HUMAN RESOURCES

A I in Human Resources is revolutionizing the way companies find, train, and manage talent. Intelligent recruitment systems can scan resumes and job applications to identify ideal candidates swiftly and without bias, making the hiring process more efficient and fairer. AI tools for employee training offer personalized learning experiences that cater to individual needs, thereby enhancing skill acquisition and job performance. Performance monitoring has also taken a leap forward with AI-driven analytics that provide real-time insights into employee productivity and engagement. By leveraging predictive tools, HR departments can foresee potential issues and proactively address them, creating a more harmonious and productive workplace. Through AI, the human element of Human Resources is not lost but rather, significantly enhanced, ensuring employees have the support and resources they need to thrive.

Intelligent Recruitment

Intelligent recruitment marks a significant departure from traditional hiring methods, interweaving artificial intelligence to streamline and enhance the entire hiring process. AI-driven recruitment isn't just about sifting through resumes faster; it's about making smarter, more informed decisions. By leveraging machine learning algorithms, predictive analytics, and natural language processing, intelligent recruitment systems can analyze vast amounts of data to identify the best candidates effectively. This technological advancement has the potential to transform how businesses find talent, saving time, reducing bias, and ensuring a better fit between company and employee.

First and foremost, the forward leap in efficiency is a compelling advantage that AI brings to recruitment. Traditionally, recruiters spend countless hours reviewing resumes, conducting initial screenings, and scheduling interviews. Intelligent recruitment systems, however, can automate these repetitive tasks, allowing HR professionals to focus on more strategic activities. By parsing resumes through sophisticated AI algorithms, these systems can rapidly identify keywords and relevant experience, providing a pre-qualification list that narrows down candidates to the most promising ones.

Moreover, one of AI's most impressive capabilities in recruitment is its ability to reduce bias. Human recruiters, despite their best intentions, can unwittingly introduce unconscious bias based on a candidate's name, gender, or educational background. Intelligent recruitment systems, however, can be designed to focus solely on the skills and qualifications that matter most for the job, thereby promoting a more inclusive hiring process. These systems don't just anonymize resumes; they contextually analyze a candidate's qualifications against job requirements to surface individuals who may have been overlooked otherwise.

In addition to minimizing bias, intelligent recruitment methods lend themselves to enhanced candidate experience. For instance, chatbots can handle initial inquiries, provide updates on application status, and even conduct preliminary interviews. Candidates often appreciate the immediate feedback and consistent communication these AI tools offer, which enhances their overall experience and perception of the company.

Predictive analytics is another transformative feature of intelligent recruitment. By analyzing historical hiring data, AI can predict which candidates are most likely to succeed in specific roles, reducing the guesswork involved in the process. For instance, if a company has identified particular personality traits and skill sets that align with high-performing employees, intelligent recruitment systems can prioritize candidates exhibiting those characteristics. This proactive approach not only aids in identifying top talent but also helps in placing these individuals in roles where they're most likely to thrive.

Natural language processing (NLP) further elevates intelligent recruitment by enabling better analysis of candidates' responses during interviews. NLP can decode the nuance in language, understanding context and sentiment, which helps in evaluating soft skills like communication, critical thinking, and problem-solving abilities. These insights can be crucial for roles that require customer interaction or team collaboration, where technical skills alone may not suffice.

Another cornerstone of intelligent recruitment is the development of custom scoring models. By tailoring the evaluation criteria to fit the unique requirements of specific roles or company cultures, these models provide a more accurate assessment of candidates. This means that rather than relying on generic metrics, recruiters can use AI to develop specialized frameworks that weigh factors that truly matter to their organization. For example, a tech startup might prioritize creativity and adaptability more than a seasoned financial institution, which may lean towards analytical rigor and reliability.

Intelligent recruitment systems can also analyze labor market trends to provide insights on talent availability, compensation benchmarks, and emerging skill sets. This information is invaluable for strategic workforce planning, enabling companies to stay competitive in attracting top talent. Recruiters can tailor their outreach and compensation packages based on real-time data, ensuring they offer attractive propositions that align with current market conditions.

Furthermore, AI-powered recruitment platforms integrate seamlessly with other HR systems, creating a cohesive ecosystem that supports the employee lifecycle from hiring to onboarding and beyond. Once a candidate is hired, these integrations help in creating personalized onboarding plans, setting initial goals, and ensuring smooth assimilation into the company culture. The cohesive flow of data ensures that new hires are well-supported from day one.

Importantly, the implementation of intelligent recruitment isn't limited to large enterprises with vast resources. AI-driven hiring tools are increasingly accessible to small and medium-sized businesses, democratizing the benefits of these advanced technologies. Cloud-based platforms offer scalable solutions that can be adapted to fit a company's size and needs, ensuring that even startups can compete in the talent acquisition game.

Lastly, as with any technology adoption, companies must be mindful of the ethical considerations surrounding AI in recruitment. While AI holds great promise in removing bias, it can also perpetuate existing biases if not properly managed. Continuous monitoring and regular updates to the algorithms are essential to ensure fairness and transparency. Companies must also be transparent with candidates about the use of AI in their hiring processes, fostering trust and clarity.

Overall, intelligent recruitment represents a paradigm shift in talent acquisition, offering unmatched efficiency, inclusivity, and accuracy. As AI technologies continue to evolve, so too will the methods we use to build our teams. Embracing intelligent recruitment can lead to faster, fairer, and more successful hiring outcomes, positioning companies to thrive in a competitive landscape.

AI in Employee Training

Artificial Intelligence (AI) is revolutionizing the way companies train their employees. No longer are traditional training workshops or lengthy manuals the only options available. AI introduces a dynamic, personalized, and 24/7 accessible training environment that meets the diverse needs of a modern workforce.

One of the game-changing elements of AI in employee training is its ability to create personalized learning experiences. Using machine learning algorithms, AI can analyze a trainee's performance in real-time and adjust the curriculum accordingly. If someone is excelling in a particular area, the AI system can offer advanced challenges to keep them engaged. Conversely, if another employee is struggling, the program can provide additional resources or alternative approaches to help them understand the material better. This custom-tailored experience ensures that every employee gets the exact training they need to succeed.

Another key feature is the use of AI-driven bots and virtual assistants to facilitate training. These technologies can provide instant feedback, answer questions, and guide employees through complex processes. Imagine working through a new software for the first time and having a virtual assistant there to guide you, offering tips and tricks along the way. This hands-on, interactive approach makes learning much more engaging compared to reading a static manual.

The adaptability of AI in employee training extends beyond individual learning. It also allows for the creation of adaptable, immersive learning environments through the use of virtual reality (VR) and augmented reality (AR). These technologies simulate real-world scenarios, providing practical experience without the associated risks. For example, in high-stakes industries like healthcare or aviation, employees can practice critical tasks in a virtual setting before applying them in the real world. This method not only enhances learning but also builds confidence and reduces errors on the job.

Data analytics is another significant advantage of AI in training programs. Employers can gather extensive data on employee performance, learning progress, and areas that need improvement. This da-ta-driven approach allows for a continual enhancement of the training process. Training modules can be updated in real-time, keeping the programs relevant and effective. Managers can also use this data to identify trends, predict future training needs, and allocate resources more efficiently.

Blended learning models that combine AI with traditional training methods are also gaining traction. By using AI to handle the routine, repetitive elements of training, human trainers are free to focus on more complex, nuanced topics that require a personal touch. This combination maximizes the strengths of both digital and human in-struction, providing a more comprehensive learning experience.

Moreover, AI-driven training can support a company's commit-ment to inclusivity and accessibility. By offering training programs that are customizable and adaptable, businesses can cater to a diverse workforce, including those with disabilities or unique learning needs. Natural language processing (NLP) and speech recognition technolo-gies can help create training modules in multiple languages, making them accessible to a global workforce.

The scalability of AI-powered training systems is another game-changer. Companies can roll out standardized training programs to thousands of employees across different locations simultaneously. There's no need to coordinate multiple trainers or sessions; the AI platform ensures that everyone receives the same high-quality training regardless of location. This capability is particularly useful for multina-tional corporations or businesses experiencing rapid growth.

The continuous improvement of voice recognition and NLP ca-pabilities has enhanced the interactivity of AI training tools. Employ-ees can engage in conversational training sessions where they interact with the system as if they were speaking to a human instructor. This interaction can be particularly beneficial for roles that require strong communication skills, such as customer service or sales. It provides a safe environment for employees to practice and refine their skills before interacting with actual clients or customers.

One of the emerging trends in AI-driven training is the use of predictive analytics to forecast future training needs. By analyzing data from various sources such as employee performance metrics, industry trends, and even market conditions, AI can predict the skills that will be in demand and suggest training programs proactively. This foresight enables companies to stay ahead of the curve, ensuring their workforce remains skilled and adaptable in a rapidly changing business environment.

AI isn't just enhancing how employees learn; it's also changing what they learn. As industries evolve, new skills and competencies become necessary. AI-driven training systems can quickly update their content to reflect these changes, ensuring that employees are always learning the most current information. This agility is crucial in sectors where innovation and technological advancement are rapid.

AI's ability to track and measure progress provides motivation and accountability for employees. Gamification elements, such as earning badges or achieving levels, can be integrated into AI training programs to make learning fun and engaging. Employees are more likely to stay committed to their training when they can see their progress and accomplishments celebrated along the way.

Furthermore, AI can foster a culture of continuous learning within an organization. Instead of viewing training as a one-time event, employees engage in ongoing education and skills development. This mindset is particularly important in today's fast-paced business world, where the ability to learn and adapt quickly often determines success.

The use and refinement of AI in employee training are still expanding, and the future holds even more potential for its application.

As AI technologies continue to evolve, they will likely become even more intuitive, making the training process smoother and more effective. It's an exciting time for both employers and employees, as AI-driven training promises to improve efficiency, engagement, and overall job performance.

AI in employee training doesn't just benefit organizations; it empowers employees. It allows them to take charge of their own learning, progress at their own pace, and identify areas where they can improve. This autonomy can lead to increased job satisfaction, better performance, and reduced turnover. When employees feel valued and invested in, they're more likely to be committed to their roles and the organization.

To summarize, the integration of AI in employee training represents a significant shift toward more personalized, efficient, and engaging educational experiences. It leverages the power of data, adaptability of machine learning, and the immersive capabilities of VR/AR to create training programs that are far superior to traditional methods. As we move forward, the continued innovation in this space suggests that the best is yet to come.

Performance Monitoring

Performance monitoring within the realm of AI in Human Resources isn't just about tracking simple metrics; it's about creating a dynamic and intuitive ecosystem that can analyze, predict, and enhance employee performance in a meaningful way. It's a blend of real-time data analysis, predictive modeling, and behavioral insights, all aimed at fostering growth and productivity in the workplace.

Firstly, AI-driven performance monitoring systems can offer real-time feedback, which is crucial for immediate improvements. Unlike traditional methods that rely on periodic reviews, AI enables continuous monitoring, providing instant feedback on various activities. Employees don't have to wait for a quarterly review to know how they're performing; they receive timely insights that help them correct course in real-time.

Moreover, AI systems can identify patterns and trends that might escape human observation. For instance, if an employee's productivity drops during certain times of the day or week, an AI system can detect this trend and suggest interventions like breaks or task adjustments. This type of nuanced analysis can lead to more personalized and effective strategies for performance enhancement.

One significant advantage of AI in performance monitoring is its ability to reduce bias. Human evaluations can often be subjective and influenced by unconscious biases. AI, on the other hand, relies on data and predefined criteria, ensuring a fairer assessment for all employees. This objectivity can contribute significantly to creating a more inclusive and equitable workplace culture.

AI also excels in predictive analytics, which can foresee potential performance issues before they become problematic. By analyzing historical data, AI can predict future performance trends and identify employees who might be at risk of underperforming. This proactive approach allows HR to intervene early, offering support or additional resources to help employees get back on track.

Additionally, AI systems can foster a culture of continuous learning and development. By identifying skills gaps and recommending targeted training programs, AI can help employees enhance their abilities and advance in their careers. This not only benefits the individual but also contributes to the overall growth and competitiveness of the organization.

For example, consider an AI system that tracks a sales team's performance. It can analyze various metrics such as sales figures, customer interactions, and time management. If the system detects a dip in performance for a particular salesperson, it can delve into the data to identify the root cause, be it a lack of follow-ups, inefficient time usage, or external market factors. The AI can then recommend specific corrective actions tailored to that individual's needs.

But it's not just about identifying weaknesses. AI can also recognize and reward high performers. By continuously scanning performance data, AI can pinpoint top achievers and suggest appropriate rewards or recognitions. This not only boosts morale but also sets a benchmark for others to strive towards.

Furthermore, AI in performance monitoring can streamline the goal-setting process. Traditional methods often involve setting annual goals that may become irrelevant as business conditions change. AI allows for dynamic goal setting, adjusting targets based on real-time data and evolving business needs. This flexibility helps ensure that employees' efforts are always aligned with the organization's strategic objectives.

Another noteworthy aspect is the role of natural language processing (NLP) in performance monitoring. NLP can analyze qualitative data from feedback forms, self-assessments, and peer reviews, extracting valuable insights that quantitative metrics alone might miss. This enables a more holistic view of an employee's performance, combining both numerical and narrative data.

Communicating performance metrics and feedback effectively is also crucial. AI-powered dashboards can present data in a visually appealing and easily understandable manner. These dashboards can be customized to highlight relevant metrics for different stakeholders, whether it's employees, managers, or executives. The intuitive design ensures that everyone has access to the information they need when they need it.

Integration with other HR systems is another key benefit of AI in performance monitoring. By seamlessly connecting with recruitment, training, and payroll systems, AI ensures a cohesive and comprehensive approach to human resource management. For instance, performance data can inform future hiring decisions, ensuring that new hires not only have the necessary skills but also align well with the company's performance culture.

Moreover, AI can facilitate more effective communication between employees and managers. Traditional performance reviews can often be stressful and anxiety-inducing. AI-driven systems can foster a more collaborative and ongoing dialogue, making the feedback process less daunting and more constructive. This continuous loop of feedback and improvement can lead to better employee engagement and satisfaction.

It's also important to consider the ethical implications of AI in performance monitoring. Transparency is crucial; employees should be aware of what data is being collected, how it is being used, and the criteria for performance evaluation. Ensuring ethical use of AI promotes trust and reduces the risk of misuse or abuse of data.

While AI offers numerous benefits for performance monitoring, it's important to remember that it should complement, not replace, human judgment. Managers still play a crucial role in interpreting data, understanding the context behind the numbers, and making informed decisions. The human touch is irreplaceable when it comes to empathy, understanding, and motivating employees.

Performance monitoring through AI is not just a futuristic concept; it's already transforming workplaces across various industries. Companies that leverage these advanced technologies position themselves to build more efficient, dynamic, and resilient workforces. By harnessing the power of AI, organizations can not only optimize performance but also foster a culture of continuous growth and development.

In conclusion, AI-driven performance monitoring offers a revolutionary approach to enhancing workplace productivity and employee development. Its ability to provide real-time feedback, identify trends and patterns, reduce bias, and offer predictive insights makes it an indispensable tool in modern human resources management. By integrating AI into performance monitoring, organizations can achieve a more fair, efficient, and motivated workforce, ultimately driving better business outcomes.

CHAPTER 21:
AI IN MANUFACTURING

In the realm of manufacturing, AI is a game-changer, revolutionizing traditional processes with unprecedented efficiency and precision. Predictive maintenance, for instance, uses AI algorithms to foresee machinery failures before they occur, significantly reducing downtime and saving costs. Robotics in manufacturing streamline assembly lines, performing repetitive and intricate tasks with flawless consistency, thereby enhancing productivity and product quality. AI's role in quality control is equally transformative; advanced vision systems can detect defects with a higher accuracy than human inspectors, ensuring that only the best products reach the market. As AI continues to integrate into manufacturing, it not only alleviates manual labor but also opens up new horizons for innovation and economic growth, making factories smarter and more agile in adapting to the ever-changing demands of today's market.

Predictive Maintenance

The concept of predictive maintenance has been a game-changer in manufacturing, thanks to the application of artificial intelligence. Predictive maintenance involves the use of advanced algorithms and machine learning models to anticipate equipment failures before they occur. This approach not only reduces downtime but also lowers maintenance costs, increases equipment lifespan, and optimizes production schedules. By leveraging vast amounts of data collected from sensors and other monitoring devices, AI algorithms can spot patterns and anomalies that human eyes might miss.

Traditional maintenance strategies often fall into two categories: reactive and preventive. Reactive maintenance, also known as "run-to-failure," involves repairing or replacing equipment only after it has failed. On the other hand, preventive maintenance schedules regular service activities based on time or usage intervals, regardless of the machine's actual condition. While preventive maintenance is certainly better than reactive, it still results in unnecessary maintenance activities and potential downtime for equipment that didn't actually need immediate attention.

Enter predictive maintenance, a more sophisticated strategy that harnesses the power of AI to predict when a machine will fail, allowing manufacturers to take action before a breakdown occurs. By analyzing historical and real-time data from sensors embedded in machinery, AI systems can identify early warning signs of potential failures. These systems can monitor a variety of factors, including temperature, vibration, sound, and other operational metrics.

One notable implementation of predictive maintenance is in the aerospace industry. Aircraft components such as engines and landing gears are equipped with sensors that continuously collect data. AI systems analyze this data to predict component wear and tear, thereby scheduling maintenance activities only when necessary. This not only ensures the safety of flights but also significantly reduces maintenance costs.

Manufacturing plants heavily rely on machinery and equipment to maintain production rates. Unexpected equipment failures can lead to significant financial losses, not just in terms of repair costs but also in lost productivity and delayed timelines. In this context, predictive maintenance offers a proactive approach. For instance, sensors on a production line may detect a slight increase in the vibration of a motor.

An AI system could interpret this as a signal that the motor is beginning to fail and schedule a maintenance check before the issue escalates.

Additionally, predictive maintenance eliminates the guesswork from maintenance schedules. Traditional preventive maintenance is based on averages and estimates, often leading to over-maintenance or under-maintenance. AI-driven predictive maintenance, however, is grounded in the actual condition of the equipment, making it more accurate and efficient. The benefits extend beyond just reducing downtime; companies can better manage their inventory of spare parts, workforce scheduling, and even extend the lifecycle of their machinery.

Incorporating predictive maintenance in manufacturing also contributes to sustainability efforts. Reducing unnecessary maintenance reduces waste, while optimizing the performance of machinery leads to more efficient energy use. In a world increasingly concerned with environmental impacts, these efficiencies can contribute to a company's sustainability goals.

The Industrial Internet of Things (IIoT) plays a crucial role in predictive maintenance by providing the infrastructure for continuous data collection. Sensors, connected through the IIoT, transmit data to a centralized system where AI algorithms process it in real time. For instance, General Electric (GE) uses its Predix platform to monitor the health of industrial machinery. This platform collects sensor data from various sources and uses machine learning models to predict equipment failures, allowing businesses to perform maintenance only when it's truly needed.

Another compelling example is the automotive industry, especially in the realm of connected cars. Modern vehicles are equipped with an array of sensors that monitor everything from engine performance to tire pressure. Predictive maintenance systems can alert drivers to potential issues before they become critical, thereby reducing the risk of costly repairs and improving vehicle safety. Companies like Tesla and BMW are already incorporating these technologies to enhance the reliability and performance of their vehicles.

Implementing predictive maintenance isn't without challenges. Companies must invest in the necessary sensor technologies and develop the infrastructure to support data collection and analysis. There is also a learning curve associated with integrating AI systems into existing maintenance processes. One practical step is to start small, perhaps by applying predictive maintenance to a single, high-value piece of equipment. Success in smaller pilot projects can build the confidence and expertise needed to scale up.

It's crucial to understand that predictive maintenance isn't just about technology; it also requires a cultural shift within organizations. Maintenance teams must become comfortable with trusting AI-driven insights and analytics. This might involve re-skilling and up-skilling the workforce to ensure that they can effectively interact with new technologies and interpret the data provided. Training programs and workshops can go a long way in easing this transition.

As AI continues to evolve, so will the capabilities of predictive maintenance. Future advancements might include the integration of augmented reality (AR) to assist maintenance workers during inspections and repairs. Imagine a technician wearing AR glasses that project real-time data and diagnostic information over the machinery, guiding them step-by-step through the maintenance process. This could significantly enhance the speed and accuracy of repairs.

Another important trend is the use of edge computing to further enhance predictive maintenance solutions. Edge computing involves processing data closer to where it's generated - at the "edge" of the network - rather than transmitting it all to a central cloud server. This can reduce latency and enable faster decision-making, which is critical in high-stakes manufacturing environments where every second counts.

Predictive maintenance powered by AI is not a distant future but an achievable present. Its applicability spans across various sectors, from automotive and aerospace to energy and pharmaceuticals. The result is a significant enhancement in operational efficiency, safety, and cost-savings. As companies continue to adopt AI technologies, predictive maintenance will undoubtedly become a standard practice, bringing us closer to a future where downtime and surprise failures are relics of the past.

In conclusion, predictive maintenance exemplifies how AI can transform traditional practices and offer tangible, real-world benefits. By adopting AI-driven predictive maintenance, manufacturers are not only making their processes more efficient but also ensuring that their operations are more reliable and cost-effective. The journey to fully integrated predictive maintenance may be challenging, but the rewards are well worth the investment, promising a future of optimized, intelligent manufacturing that works proactively rather than reactively.

Robotics in Manufacturing

When discussing "Robotics in Manufacturing," one might visualize advanced machines undertaking complex tasks with precision and efficiency. This visualization is not far from reality. The integration of robotics in manufacturing is revolutionizing industries by enhancing productivity, ensuring quality, and reducing costs. These sophisticated robots, powered by AI, are capable of performing repetitive tasks, handling materials, and assembling products with incredible accuracy. The evolution of robotic technology is steering us toward a future where manufacturing processes are not only faster but also more adaptable to changing demands.

AI-powered robots are equipped with sensors and machine learning algorithms that enable them to navigate and make decisions autonomously. This level of autonomy mitigates the need for human intervention, resulting in a significant reduction in human error. Manufacturers can rely on these robots to maintain consistent output quality, leading to fewer defective products. This consistency is particularly crucial in industries where precision is paramount, such as automotive and electronics manufacturing.

The industrial robots of today are far more sophisticated than their predecessors. They are no longer confined to cages or restricted to simple, repetitive tasks. Collaborative robots, or "cobots," can work alongside human operators in a shared workspace. These cobots are designed with advanced safety features, ensuring that they can detect and respond to human presence, minimizing the risk of accidents. This human-robot collaboration opens new avenues for enhancing workplace efficiency and safety. Workers can focus on complex, cognitive tasks while robots handle the more mundane, physically demanding ones.

Moreover, robotics is not just limited to the actual manufacturing process. Robots are increasingly being used for inventory management and logistics within the factory floor. These robots can move raw materials and finished products between different stations, ensuring a seamless flow of production. By integrating AI, these robots can optimize their routes, perform real-time inventory tracking, and even predict future stock requirements. This level of automation reduces downtime and improves overall supply chain efficiency.

The implementation of robotics in manufacturing also has significant implications for scalability. As businesses grow, the demand for increased production capacity arises. Adding more robots to the production line is much easier and faster than training new human workers. Robots can be programmed to perform new tasks with minimal downtime, allowing manufacturers to quickly adapt to changing market demands. This flexibility is a competitive advantage in an ever-evolving market landscape.

While the benefits are manifold, the integration of robotics in manufacturing does come with its challenges. One major concern is the initial capital investment required for procuring and implementing robotic systems. However, the long-term return on investment often outweighs these initial costs, as robotics leads to substantial savings in labor costs, reduced waste, and increased production efficiency. Manufacturers must carefully evaluate their business needs and conduct a cost-benefit analysis before embarking on the journey of robotic integration.

Another significant challenge is the need for skilled personnel to operate and maintain these advanced robotic systems. This has led to a growing emphasis on upskilling the workforce. Educational institutions and training programs are starting to offer courses specifically focused on robotics and AI. By equipping the workforce with the necessary skills, businesses can ensure a smoother transition to automated manufacturing processes. This shift in skill requirements also highlights the importance of lifelong learning and continuous professional development in the modern workforce.

In addition to these challenges, ethical considerations also come into play. The rise of robotics in manufacturing raises questions about job displacement and the socioeconomic impact on the workforce. While robots can undoubtedly perform many tasks more efficiently than humans, the transition must be managed carefully to mitigate potential negative effects on employment. Policymakers and industry leaders must work together to create strategies that promote a balanced approach, ensuring that the benefits of automation are shared across society.

From an environmental perspective, the efficiency gained through robotics can also contribute to sustainability efforts. Robots can optimize the use of resources, reduce waste, and minimize energy consumption. For instance, precision in material handling and assembly can lead to less scrap and rework, lowering the overall environmental footprint of the manufacturing process. Robotics, when integrated intelligently, can be a cornerstone of sustainable manufacturing practices.

The future of robotics in manufacturing is brimming with potential. Emerging technologies such as cloud computing, edge computing, and the Internet of Things (IoT) are further enhancing the capabilities of robotic systems. These technologies enable robots to process data in real-time, communicate with other machines, and even learn from each other. This interconnected ecosystem, often referred to as Industry 4.0, represents the next frontier in manufacturing innovation.

Imagine a factory where every machine, robot, and system is connected and constantly sharing data. This level of integration allows for unprecedented levels of efficiency, predictive maintenance, and intelligent decision-making. If a machine detects a potential fault, it can automatically adjust its operations or even request maintenance before a failure occurs. This proactive approach minimizes downtime and ensures uninterrupted production.

The possibilities don't end there. Advances in artificial intelligence and machine learning are paving the way for even more advanced autonomous robots. These robots will be capable of learning new tasks through observation and imitation, much like humans do. This kind of adaptability will be a game-changer, allowing robots to perform a wide range of tasks without the need for reprogramming. It will also make the transition to automated systems smoother for industries that deal with a high variety of products and processes.

In conclusion, the integration of robotics in manufacturing is transforming the industry in ways that were once considered science fiction. From improving efficiency and product quality to promoting workplace safety and sustainability, robotics powered by AI is setting new standards for what is achievable in manufacturing. While challenges exist, they are not insurmountable, and the benefits far outweigh the drawbacks. As technology continues to advance, the future of manufacturing will undoubtedly become even more dynamic, efficient, and sustainable.

This transformation is not just about the technology itself but about how we harness it to create better, more efficient, and sustainable manufacturing processes. The journey is ongoing, and we are only at the beginning of realizing the full potential of robotics in manufacturing. It is an exciting time for the industry, and those who embrace this change will be well-positioned to thrive in the future.

Quality Control

In the realm of manufacturing, quality control is crucial for ensuring that products meet specifications and customer expectations. Traditional methods have relied heavily on human inspection and consistent manual processes. However, as technology advances, Artificial Intelligence (AI) offers more efficient, accurate, and scalable quality control solutions.

AI in quality control involves using machine learning algorithms and computer vision systems to inspect products and identify defects or inconsistencies in real time. These systems are capable of analyzing vast amounts of data at speeds no human could match, significantly enhancing the precision and efficiency of the quality control process. By leveraging the power of AI, manufacturers can dramatically reduce the incidence of defective products making it to the consumer.

Machines equipped with computer vision can scrutinize products faster and more accurately than human inspectors. Cameras and sensors capture images of the products, which are then analyzed by AI algorithms trained to detect defects. These defects could be anything from minor cosmetic imperfections to critical faults that could affect the functionality or safety of the product. For instance, in the automotive industry, AI can inspect car parts for defects, ensuring that only those meeting strict quality standards are used in assembly.

AI systems also offer predictive capabilities. They can analyze patterns from historical data to predict potential defects before they occur. This foresight allows manufacturers to take preemptive action, addressing issues in the production line before they lead to faulty products. Predictive quality control minimizes wasted materials and resources, ultimately reducing costs while maintaining high standards of production.

Moreover, AI can be integrated into different stages of the production process. From inspecting raw materials at the onset to examining finished products before they are shipped, AI provides a continuous quality-check loop. This end-to-end quality control ensures that issues are caught as early as possible, minimizing the need for rework and ensuring that finished products consistently meet quality expectations.

One of the significant advantages of AI-driven quality control is its ability to learn and adapt over time. Machine learning algorithms can learn from each inspection, continuously improving their accuracy and capability. This adaptive learning means that the system gets better at identifying subtle defects that might have been overlooked previously. The agility of AI systems contrasts sharply with the challenges of retraining human inspectors or reconfiguring traditional inspection machines.

In addition to their precision and adaptability, AI-driven quality control systems offer unparalleled scalability. Whether a manufacturer is producing a few thousand units or scaling up to millions, AI systems can handle the increased workload without compromising on inspection speed or accuracy. This scalability is particularly beneficial for industries facing fluctuating demand, enabling them to maintain quality control standards regardless of production volume.

AI has also significantly reduced the response time between defect detection and corrective action. In a traditional setup, identifying a defect might involve multiple steps, including manual inspection, reporting, and decision-making on corrective measures. AI systems cut through this bureaucratic lag by instantly notifying relevant personnel and even triggering automated responses to address defects. This rapid response capability helps in maintaining a seamless production flow, minimizing down-times and disruptions.

Besides identifying defects, AI-powered quality control can also provide insights into process improvements. By analyzing data from production lines, AI can identify trends and anomalies that suggest inefficiencies. These insights enable manufacturers to optimize their processes, reducing waste and ensuring that operations run smoothly. The result is not just higher quality products but also more efficient and cost-effective production methods.

Furthermore, AI quality control systems contribute to workforce safety. In many manufacturing environments, quality control inspections can expose workers to hazardous conditions. AI can take over these risky tasks, reducing human exposure to dangerous elements. Consequently, this not only ensures that quality checks are thorough and consistent but also that workers are safer and can focus on less perilous aspects of their roles.

Facilitating remote quality control is another notable application of AI in this context. Through IoT (Internet of Things) connectivity, AI systems can be monitored and operated remotely. Manufacturers can deploy AI-driven quality control systems across multiple geographical locations and oversee their operations from a central hub. This centralization simplifies management, reduces travel costs, and ensures uniform quality standards across different production sites.

Moreover, AI aids in documenting and maintaining quality records. Automated inspections generate a comprehensive data trail that includes images, defect logs, and corrective actions. This documentation is invaluable for compliance with industry regulations and standards. It also helps manufacturers during audits and when providing evidence of quality assurance to clients or stakeholders.

As AI technology continuously evolves, its application in quality control will become more sophisticated. Deep learning, for instance, can further boost defect detection capabilities by recognizing complex patterns that simpler algorithms might miss. The integration of AI with other emerging technologies such as augmented reality (AR) and virtual reality (VR) can also enhance quality inspection processes by providing immersive and interactive inspection tools.

In summary, the use of AI in quality control represents a significant advancement over traditional methods. It offers unmatched accuracy, speed, and scalability, while also providing predictive insights and ensuring worker safety. By embedding AI into the quality control processes, manufacturers can not only enhance product quality but also achieve cost efficiencies and operational optimization. As we move forward, the potential applications of AI in this space are bound to expand, further revolutionizing how quality control is approached in the manufacturing sector.

CHAPTER 22:
AI IN LOGISTICS AND SUPPLY CHAIN

In an era where efficiency is paramount, AI's integration into logistics and supply chain management is a game-changer that optimizes operations from start to finish. By leveraging intelligent inventory management systems, businesses can ensure optimal stock levels and reduce waste. Route optimization algorithms meticulously plan the most efficient delivery paths, saving time and fuel costs while enhancing customer satisfaction. AI in demand forecasting transforms data into actionable insights, allowing companies to predict market trends and consumer needs with unprecedented accuracy. This convergence of AI technologies creates a streamlined, responsive supply chain that can adapt quickly to disruptions, driving both profitability and sustainability in an ever-evolving market landscape.

Intelligent Inventory Management

One of the cornerstones of any successful logistics and supply chain operation is effective inventory management. Historically, inventory management has relied heavily on human oversight and decision-making, involving a considerable amount of guesswork and manual processes. But with the advent of Artificial Intelligence (AI), these traditional methods are undergoing a significant transformation. Intelligent inventory management solutions powered by AI are not just improving efficiency but also paving the way for a more dynamic and responsive supply chain.

At its core, intelligent inventory management leverages AI algorithms to analyze vast amounts of data in real-time. From sales patterns and seasonal trends to customer behaviors and supplier performance, AI can process and interpret complex datasets far more quickly and accurately than a human ever could. By continuously refining these data points, AI systems can offer predictive insights that help businesses maintain optimal inventory levels.

AI-driven inventory management systems are capable of minimizing excess stock and reducing stockouts, both of which can be costly for businesses. Excess inventory ties up capital and increases storage costs, while stockouts lead to missed sales opportunities and can erode customer loyalty. The balance between having too much and too little inventory is delicate, but AI can forecast demand with high accuracy, making this balancing act much easier.

Imagine a scenario where a retailer uses AI to predict that the demand for winter coats will increase due to an upcoming cold front. The AI system automatically sends replenishment orders to suppliers, ensuring that the retailer has ample stock to meet customer demand without overstocking. This proactive approach contrasts sharply with the reactive methodologies of traditional systems, which often respond only after inventory levels have reached critical lows or highs.

One way AI achieves this is through machine learning algorithms that continually improve over time. Initially, these algorithms may use historical data to make predictions. However, as new data flows in, the system learns from past mistakes and successes to fine-tune its predictions. This iterative process leads to increasingly accurate forecasts, which can substantially improve a company's bottom line.

Furthermore, AI-powered inventory management systems are not limited to predicting demand. They also take into account supply-side variables. For example, if a supplier has a history of late deliveries, the AI system can adjust its order schedules accordingly or even suggest alternate suppliers. This ensures that unexpected disruptions have minimal impact on the inventory levels, keeping the supply chain running smoothly.

Another critical application of AI in inventory management is the optimization of warehouse operations. Traditional warehouse management involves labor-intensive tasks such as picking, packing, and shipping products. AI simplifies these processes by providing real-time insights into warehouse operations, identifying inefficiencies, and suggesting improvements. For example, AI can recommend the optimal layout for storing products based on their turnover rates, which can significantly speed up the picking process.

The use of AI in inventory management also extends to improving transparency and visibility within the supply chain. Blockchain technology, when combined with AI, can offer a tamper-proof and transparent ledger of inventory movements. Every transaction or movement is recorded and validated, providing an end-to-end overview of the supply chain. This is particularly beneficial in sectors where traceability and compliance are critical, such as pharmaceuticals and food industries.

AI's role doesn't stop at optimizing what is already in place; it also helps in anticipating future trends that can impact inventory management. For example, consumer preferences are always evolving. With AI, businesses can analyze social media trends, customer reviews, and other unstructured data to forecast how these preferences might translate into demand. This level of foresight can be invaluable for product launches or seasonal promotions.

Chatbots and virtual assistants are other AI tools transforming inventory management. These AI applications can handle customer service inquiries related to product availability, delivery status, or return processes. By automating these routine interactions, businesses free up human employees to focus on more complex tasks, thereby improving operational efficiency.

Intelligent inventory management also offers environmental benefits. By accurately forecasting demand and optimizing stock levels, businesses can reduce waste. This is particularly important in industries dealing with perishable goods. Efficient inventory management ensures that less product goes unsold and ends up in landfills, contributing positively to sustainability goals.

Moreover, AI solutions provide real-time analytics and reporting, offering businesses the capability to make data-driven decisions instantly. Dashboards that display key performance indicators (KPIs) related to inventory levels, turnover rates, and order accuracy can help managers identify trends and spot issues before they become significant problems. This level of insight is crucial for maintaining a competitive edge in today's fast-paced market landscape.

Adopting intelligent inventory management systems isn't without its challenges. Integrating these advanced systems with existing ERP (Enterprise Resource Planning) solutions requires careful planning and robust IT support. Additionally, there's the initial investment in both technology and training. However, the long-term benefits—ranging from cost savings and increased efficiency to improved customer satisfaction—far outweigh the upfront costs.

Moreover, companies must be mindful of the ethical considerations tied to AI implementation. Issues such as data privacy and algorithmic bias need careful monitoring. Transparency in how AI systems make decisions can build trust both within the organization and with its customers. Ensuring ethical AI usage aligns with broader corporate social responsibility goals and regulations.

Lastly, the potential for AI-driven inventory management to evolve is immense. As technology advances, so will the capabilities of these systems. Future innovations could include more sophisticated predictability models, enhanced real-time collaboration tools, and even greater integration with other AI systems across the entire supply chain. For instance, coupling intelligent inventory management with AI-powered transportation logistics could revolutionize how goods are moved, stored, and delivered.

In summary, intelligent inventory management is a powerful application of AI that offers numerous advantages—from enhanced efficiency and cost savings to improved sustainability. While implementation challenges exist, the significant long-term benefits make it a worthwhile investment. As AI technology continues to advance, the potential for intelligent inventory management to transform logistics and supply chain operations grows, heralding a new era of excellence and innovation in the industry.

Route Optimization

In the dynamic field of logistics and supply chain management, route optimization stands as a critical application of artificial intelligence. By leveraging advanced algorithms and vast datasets, AI drives efficiencies that were previously unattainable. This not only reduces costs but also enhances service levels, ultimately contributing to a more seamless consumer experience.

One of the fundamental advantages of AI-driven route optimization is its ability to process and analyze a myriad of variables. Unlike traditional route planning methods that might consider just the distance between points, AI can evaluate factors such as traffic patterns, weather conditions, vehicle capacity, and delivery time windows. This multifaceted approach ensures that each route is tailored for maximum efficiency and effectiveness.

Imagine a retail company with a fleet of delivery trucks. Using AI for route optimization, the company can dynamically adjust routes in real-time based on current traffic conditions, thereby avoiding congested areas and reducing delivery times. This not only saves fuel costs but also enhances customer satisfaction by ensuring timely deliveries.

Furthermore, AI-based route optimization allows for dynamic rerouting. If an unexpected event, such as an accident or road closure, occurs, the system can quickly recalculate the best possible route. This level of adaptability is invaluable in a world where time-sensitive deliveries can make or break a business. By ensuring that goods are delivered as scheduled, companies can uphold their commitments and build stronger relationships with their customers.

Another exciting application of AI in route optimization is the integration with autonomous vehicles. As self-driving technology continues to evolve, AI systems can coordinate fleets of driverless trucks, planning and optimizing routes with unparalleled precision. These autonomous vehicles can operate around the clock, breaking the constraints of human fatigue and driving regulations, thus further increasing the efficiency of logistics operations.

Besides transportation, AI-driven route optimization also has significant implications for warehouse logistics. In large warehousing environments, it can be used to streamline the movement of goods within the facility. Automated guided vehicles (AGVs) and robots can be directed through the most efficient paths to pick and place items, reducing the time taken to fulfill orders and minimizing the risk of bottlenecks.

The environmental benefits of AI in route optimization should not be overlooked either. By minimizing unnecessary travel and maximizing load efficiency, companies can significantly reduce their carbon footprint. Route optimization contributes to more sustainable operations, aligning with the growing emphasis on environmental responsibility and conservation in the logistics industry.

Moreover, AI can continuously improve route planning by learning from historical data. Machine learning algorithms analyze past delivery data to identify patterns and trends that can inform future route optimizations. This continuous learning loop means that the system gets smarter and more efficient over time, offering incremental improvements that compound into significant operational gains.

Consider the potential of predictive analytics in this context. By forecasting demand and anticipating potential disruptions, AI can proactively plan routes that mitigate risks before they even materialize. This forward-thinking approach transforms logistics from a reactive to a proactive discipline, offering a strategic advantage in a highly competitive market.

In the realm of supply chain visibility, AI-enhanced route optimization plays a crucial role. With end-to-end tracking of shipments and real-time alerts, stakeholders can have a clear view of where their goods are at any given moment. This transparency allows for better coordination and communication across the supply chain, reducing delays and increasing accountability.

Furthermore, AI systems can integrate with other technologies such as the Internet of Things (IoT) to collect real-time data from sensors embedded in vehicles and infrastructure. This symbiotic relationship between AI and IoT ensures that the most current information feeds into the route optimization process, leading to more accurate and timely decisions.

Smaller businesses stand to gain substantially from AI-driven route optimization as well. With the availability of AI as a service, even companies without extensive technical resources can access sophisticated optimization tools. This democratization of technology levels the playing field, allowing smaller players to compete more effectively in the logistics arena.

In summary, AI-driven route optimization is revolutionizing logistics and supply chain management by enhancing efficiency, reducing operational costs, increasing customer satisfaction, and contributing to environmental sustainability. By adopting these advanced technologies, organizations can expect to see significant improvements in their delivery capabilities and overall operational performance.

AI in Demand Forecasting

Demand forecasting is the bedrock of a functional and efficient supply chain. Getting accurate forecasts means a smoother operation, less wasted inventory, and happier customers. Traditionally, demand forecasting relied on historical data and statistical methods. These methods brought value but also had their limitations—especially in dealing with complex variables and rapid market changes. This is where Artificial Intelligence (AI) steps in to offer a revolutionary approach, blending data analytics with cognitive computing for a more precise and adaptive forecasting model.

Artificial Intelligence makes its mark in demand forecasting by analyzing vast amounts of data more quickly and accurately than any human could. It employs machine learning algorithms to identify patterns and trends that are not immediately apparent through traditional methods. By drawing from multiple data sources—such as historical sales, current market trends, weather conditions, and even social media activity—AI can create a detailed and dynamic forecast that adjusts in real-time.

At its core, AI employs technologies such as neural networks, deep learning, and natural language processing to parse through data. Neural networks are particularly effective in spotting intricate patterns and relationships within large datasets. These algorithms learn and improve over time, continually fine-tuning their predictive capabilities. The more data the system processes, the smarter and more accurate it becomes.

Consider this: A retail chain using AI for demand forecasting can integrate data from point-of-sale systems, supply chain logistics, and even local events. A sudden spike in demand because of an upcoming festival can be anticipated and accounted for, ensuring that stock levels meet customer requirements without overstocking. This sort of nuanced forecasting helps in aligning supply with demand more fluidly and efficiently.

AI in demand forecasting doesn't just stop at predicting sales numbers. It creates a comprehensive reflection of market sentiments, consumer behaviors, and potential disruptors. For instance, AI can predict a product's lifecycle by analyzing its emerging trends, seasonal variations, and even competitor actions. Companies can use these forecasts to decide when to launch new products, phase out old ones, or prepare for promotional periods.

Moreover, one of the most significant advantages AI brings to demand forecasting is its ability to mitigate risks associated with market volatility. Business landscapes today can change at the blink of an eye due to geopolitical shifts, economic fluctuations, or even sudden changes in consumer preferences. AI enables businesses to adapt to these changes swiftly. Machine learning algorithms can rapidly assimilate new data points and adjust forecasts accordingly. This adaptability ensures that businesses remain agile, proactive, and prepared for any sudden changes in the demand curve.

Another essential factor is cost efficiency. Over-forecasting can lead to excessive inventory, tying up capital that could be better used elsewhere. Under-forecasting, on the other hand, can lead to stockouts and lost sales. AI minimizes these financial pitfalls by enhancing forecast accuracy. As a result, businesses can optimize their inventory levels, reduce storage costs, and improve cash flow management.

One real-world example is how Amazon uses AI in its supply chain. The e-commerce giant leverages machine learning algorithms to analyze shopping patterns, weather data, and even traffic conditions to predict product demand with astounding accuracy. This precise forecasting allows Amazon to optimize its inventory levels, ensuring that products are available when and where customers need them, while minimizing the costs associated with excess inventory.

Additionally, AI in demand forecasting is not limited to large corporations alone. Small and medium-sized enterprises (SMEs) can also harness the power of AI to improve their demand forecasting processes. With the availability of cloud-based AI tools and platforms, SMEs can access advanced forecasting capabilities without the need for substantial in-house IT infrastructure. This democratization of technology enables businesses of all sizes to compete on a more level playing field.

Implementing AI in demand forecasting isn't without its challenges. One primary concern is data quality. For AI algorithms to function optimally, they require clean, accurate, and comprehensive data sets. Integrating data from disparate sources and ensuring its reliability can be a complex task. Organizations must invest in data management and governance practices to address this issue effectively.

Another challenge is the ethical use of AI. As AI systems analyze vast amounts of data, including consumer behavior and personal information, privacy concerns come to the forefront. It is crucial for organizations to adhere to ethical standards and regulatory requirements when deploying AI-driven demand forecasting solutions. Transparency and data security are paramount in building trust with customers and stakeholders.

Despite these challenges, the benefits of AI in demand forecasting are undeniable. Businesses that embrace AI-driven forecasting solutions gain a competitive edge in the market. They can respond more swiftly to changes in demand, reduce operational costs, and enhance customer satisfaction. AI empowers organizations to make data-driven decisions that drive growth and profitability.

In summary, AI in demand forecasting marks a paradigm shift in supply chain management. By harnessing the power of machine learning, neural networks, and data analytics, businesses can achieve unparalleled accuracy and agility in predicting demand. This transformative technology is reshaping industries, enhancing efficiency, and ultimately improving the way we live and work. As AI continues to evolve, its role in demand forecasting will only become more critical, ensuring that businesses are better prepared for the challenges and opportunities of the future.

CHAPTER 23:
AI IN ENERGY AND UTILITIES

In the realm of energy and utilities, AI is transforming how we produce, distribute, and consume power by bringing unprecedented
efficiency and sustainability. Imagine smart grids that can balance supply and demand in real-time, reducing operational costs and ensuring
more reliable service. AI algorithms analyze vast amounts of data to
forecast energy needs accurately, allowing for better resource management and reducing wastage. Additionally, intelligent systems can allocate resources more effectively, optimizing the usage of renewable energy sources like wind and solar power. This not only helps in cutting
down carbon emissions but also in managing the natural resources
more judiciously. The integration of AI in energy and utilities is not
just a technological upgrade; it's a giant leap toward a more sustainable
and economically sound future, making a real difference in how we
power our lives daily.

Smart Grids

Smart grids represent a significant breakthrough in the realm of energy
and utilities, leveraging artificial intelligence to create more efficient,
reliable, and sustainable energy systems. Unlike traditional grids, which
operate on a one-way system where electricity flows from providers to
consumers, smart grids facilitate a two-way dialogue. This
bi-directional communication is pivotal in integrating renewable

energy sources, enhancing grid stability, and optimizing power distribution.

The core of a smart grid is its ability to collect and analyze vast amounts of data in real-time. Advanced metering infrastructure (AMI), sensors, and IoT devices gather a multitude of data points, ranging from electricity consumption patterns to the health of grid components. AI algorithms then process this data to predict demand, identify potential issues before they escalate, and dynamically adjust the grid's performance. For instance, during periods of high demand, AI-driven systems can reroute electricity to prevent overloads and outages, ensuring a seamless supply of power to consumers.

One of the primary benefits of smart grids is the efficient integration of renewable energy sources like solar and wind. Traditional grids struggle with the intermittent nature of these energy sources, often leading to either a surplus or shortage of power. AI can forecast weather conditions and predict energy yield, allowing the grid to balance supply and demand more efficiently. For example, if a sunny day is projected, the AI can anticipate a higher output from solar panels and adjust other energy resources accordingly. This not only maximizes the use of renewables but also reduces dependency on fossil fuels, contributing to environmental sustainability.

Beyond managing electricity flow, smart grids contribute significantly to operational efficiency and maintenance. Predictive maintenance powered by AI can foresee potential equipment failures by analyzing patterns and anomalies in the data. This helps utility companies perform proactive maintenance, reducing downtime and extending the lifespan of infrastructure components. Instead of dealing with unexpected blackouts or equipment failures, utility companies can address issues before they become critical, resulting in cost savings and improved service reliability.

Consumer engagement also sees a transformation with the advent of smart grids. Traditional energy bills provide a retrospective view of consumption, leaving consumers with little control over their energy use. Smart grids, however, empower consumers with real-time data. Mobile apps and web portals offer insights into energy consumption, allowing users to monitor their usage patterns and make informed decisions. Whether it's adjusting the thermostat when energy prices peak or scheduling heavy-duty appliances for off-peak hours, these small changes can cumulatively lead to significant savings.

Additionally, dynamic pricing models are enabled through smart grids. Instead of a fixed rate, dynamic pricing fluctuates based on real-time supply and demand. When demand is low, prices drop, encouraging consumers to use energy. Conversely, during peak times, prices rise, incentivizing reduced consumption. This pricing strategy not only helps in balancing the grid load but also fosters a more efficient and sustainable energy use culture among consumers.

The security implications of smart grids cannot be understated. With more data and interconnected devices, the potential for cybersecurity risks increases. AI plays a critical role in safeguarding these systems. Advanced machine learning algorithms can detect and thwart cyber threats by recognizing unusual patterns that human operators might overlook. From identifying unauthorized access to preventing data breaches, AI ensures that the integrity of the grid remains intact, creating a resilient defense mechanism against cyber-attacks.

Furthermore, smart grids facilitate better disaster response and recovery. In the event of a natural disaster or unforeseen breakdown, AI can quickly identify the affected areas and reroute power from unaffected zones. This rapid response minimizes downtime and ensures that critical services like hospitals and emergency response centers remain operational. The ability to dynamically reconfigure the grid enhances its resilience, making communities better equipped to handle crises.

The transition to smart grids also spurs economic benefits. The implementation of these advanced systems creates job opportunities in various sectors including data science, cybersecurity, and engineering. Training programs and educational courses focusing on AI and smart grid technologies emerge, fostering a skilled workforce ready to tackle the challenges of modern energy systems. Moreover, the cost savings achieved through more efficient energy distribution and reduced maintenance translate to lower utility bills for consumers in the long run.

It's not just about technology; policy and regulatory frameworks play a crucial role in the widespread adoption of smart grids. Governments and regulatory bodies must provide the necessary support through incentives, grants, and favorable policies. Collaboration between public and private sectors can accelerate research and development, paving the way for innovation in smart grid technologies. Countries that invest in smart grids are likely to see long-term economic and environmental dividends, positioning themselves as leaders in sustainable energy management.

In conclusion, smart grids, powered by artificial intelligence, are redefining the landscape of energy and utilities. By enabling efficient integration of renewable energy, enhancing grid reliability, empowering consumers, and ensuring robust security, smart grids are paving the way for a more sustainable and efficient energy future. Embracing this technology means not just optimizing energy distribution, but also securing our planet's well-being for generations to come.

AI in Energy Forecasting

Artificial Intelligence (AI) is revolutionizing various industries, and the energy sector is no exception. One of the most transformative applications of AI in this realm is energy forecasting. Accurate energy forecasting is crucial for balancing supply and demand, optimizing energy usage, and reducing costs for both providers and consumers. By leveraging AI, the energy sector can enhance its forecasting capabilities, leading to more efficient and sustainable energy management.

Traditional energy forecasting methods have relied heavily on human expertise and historical data. While these methods have served well in the past, they often lack the precision and adaptability required to meet the dynamic needs of today's energy landscape. AI, on the other hand, offers a data-driven, adaptable approach that can handle complex and fluctuating variables. This ability to process vast amounts of data in real time makes AI an invaluable tool for energy forecasting.

Machine learning algorithms, a subset of AI, have shown remarkable promise in improving the accuracy of energy forecasts. These algorithms can analyze a variety of data sources such as weather patterns, historical energy consumption, and real-time grid data to predict future energy needs. For instance, by analyzing weather data, AI can forecast the amount of solar or wind energy that will be generated, helping utility companies plan accordingly.

Incorporating AI into energy forecasting can lead to significant cost savings and operational efficiencies. For providers, accurate forecasts mean they can better predict peak demand times and adjust energy generation accordingly. This reduces the need for expensive and often environmentally harmful backup generators. For consumers, accurate forecasting translates to more stable energy prices and fewer disruptions in supply.

Moreover, AI in energy forecasting can contribute to environmental sustainability. By optimizing energy generation and consumption, AI helps reduce waste and lower carbon emissions. For example, if a utility company can forecast a surplus of renewable energy, they can distribute it efficiently, minimizing the reliance on fossil fuels. This not only makes economic sense but also aligns with global efforts to combat climate change.

The advent of smart grids further enhances the capabilities of AI in energy forecasting. Smart grids use AI to gather real-time data from various points in the energy distribution network. This data-driven approach allows for more precise and responsive energy management, ensuring that energy is available where and when it is needed. By integrating AI with smart grids, utility companies can achieve unprecedented levels of efficiency and reliability.

Advanced AI models also facilitate demand response programs, which adjust energy consumption based on supply conditions. These programs are particularly beneficial during periods of high demand or limited supply, helping to prevent blackouts and reduce energy costs. AI can predict when these periods are likely to occur and recommend adjustments in energy usage, making it easier for both providers and consumers to adapt.

However, the implementation of AI in energy forecasting is not without challenges. One of the primary concerns is the quality and availability of data. Accurate forecasting requires high-quality, real-time data, which can be difficult to obtain. Additionally, integrating AI systems with existing infrastructure requires significant investment and coordination among various stakeholders.

Despite these challenges, the potential benefits of AI in energy forecasting are too significant to ignore. As technology continues to advance, it is likely that the accuracy and reliability of AI-driven forecasts will only improve. This will pave the way for a more efficient, cost-effective, and sustainable energy sector.

One notable case study is the application of AI in the energy markets of Europe. Several countries have already begun integrating AI into their energy forecasting models with impressive results. For instance, AI-driven forecasting has helped European grid operators reduce the margin of error in their predictions, leading to better resource management and lower operational costs.

Another promising development is the use of deep learning techniques in energy forecasting. Deep learning, a more complex subset of machine learning, involves neural networks with many layers that can model intricate patterns in data. These techniques have the potential to further enhance the precision of energy forecasts, particularly in scenarios with highly variable and complex data.

Additionally, collaborative efforts between academia, industry, and government agencies are driving innovation in this field. Research institutions are developing new algorithms and models to improve forecasting accuracy, while industry players are investing in the infrastructure needed to implement these advancements. Government agencies are also playing a crucial role by supporting initiatives and providing regulatory frameworks that facilitate the adoption of AI in the energy sector.

In conclusion, AI is set to revolutionize energy forecasting by offering more accurate, efficient, and sustainable solutions. By harnessing the power of AI, the energy sector can better manage supply and demand, optimize resource utilization, and contribute to environmental sustainability. The journey toward fully realizing the potential of AI in energy forecasting may be complex and challenging, but the rewards are well worth the effort. With continued investment and innovation, AI will undoubtedly play a pivotal role in shaping the future of energy management.

Intelligent Resource Allocation

The use of Artificial Intelligence (AI) in energy and utilities is dramatically transforming the landscape of resource management. One of the most revolutionary impacts is seen in intelligent resource allocation.

This refers to the optimal distribution and utilization of resources like electricity, water, and gas, ensuring that these essentials are both available and efficient.

Traditionally, resource allocation has been a complex and often inefficient process managed by human operators relying on historic data and approximations. With AI, we're shifting from a reactive model to a proactive one, where advanced algorithms predict demand and optimize supply chains in real-time. This enables utility companies to anticipate peaks and troughs in usage, reducing waste and preventing shortages.

Machine learning algorithms play a crucial role in forecasting. These models analyze vast amounts of historical data to identify patterns and relationships that might not be immediately apparent to human analysts. For example, by studying energy consumption trends alongside weather patterns, AI can predict spikes in electricity demand before a significant weather change occurs. The algorithms continually improve, learning from new data to make increasingly accurate forecasts.

Moreover, AI-driven resource allocation isn't just about prediction; it also facilitates real-time decision-making. This is accomplished through various IoT devices and smart meters that provide continuous data streams. These devices, embedded across the grid and within individual homes, communicate with central management systems. AI processes this data to dynamically distribute resources, adjusting for any anomalies or unexpected demand surges almost instantaneously.

Consider a scenario where a sudden heatwave causes residential demand for electricity to skyrocket. An AI system can reallocate resources by momentarily diverting energy from less critical areas or throttling energy-intensive industrial operations. This level of response wasn't possible with legacy systems but is becoming standard with AI integration. Such automation ensures a balanced grid, prevents black-outs, and maintains service reliability.

Additionally, AI enhances the efficiency of renewable energy sources like solar and wind. These energy types are inherently variable, dependent on weather conditions that fluctuate daily and seasonally. By using AI to predict weather conditions and integrate these predictions into resource planning, utility companies can optimize the distribution of renewable energies, thereby reducing reliance on fossil fuels and contributing to environmental sustainability.

Cost-efficiency is another significant benefit. Efficient resource allocation drastically reduces operational costs for utilities. By minimizing waste and optimizing resource distribution, companies can cut down on the expenses associated with excessive production and storage. Furthermore, AI enables predictive maintenance, identifying equipment that might soon fail and require repair or replacement. This preemptive action prevents costly outages and extensive repair jobs, resulting in both economic and operational advantages.

Now, think about the environmental implications. Reducing waste and optimizing renewable energy use directly contribute to a cleaner, greener planet. AI's ability to manage resources intelligently means fewer carbon emissions, less environmental degradation, and a more sustainable future. By streamlining these resources, utility companies not only meet demand more effectively but also play a vital role in environmental stewardship.

Moreover, AI-powered systems can integrate consumer behavior into their models. Smart thermostats, smart appliances, and other home automation devices provide valuable data on how individuals use energy. This information allows for micro-level adjustments that enhance overall resource allocation. For instance, if a majority of households in a neighborhood are lowering their thermostats during a particular time, the system can reduce the overall allocation of energy to that area without impacting comfort, thereby saving resources.

AI also brings democratization of energy management with decentralized energy resources. Consumers who generate their own energy through solar panels, for instance, can become 'prosumers', feeding surplus energy back into the grid. AI manages and optimizes this bidirectional flow of energy, ensuring that both providers and consumers benefit from the exchange. The result is a more dynamic and resilient energy system.

It's also essential to consider the safety and security aspects. Intelligence in resource allocation includes robust mechanisms for detecting and mitigating risks. AI can identify irregular patterns that might indicate a leak, fault, or cyber intrusion. Early detection systems enable operators to take swift corrective action, thus safeguarding both the infrastructure and the end-users.

AI's impact on intelligent resource allocation also extends to strategic planning. By offering insights based on comprehensive data analysis, AI helps utility companies make informed decisions about infrastructure investments, policy changes, and future resource needs. This long-term vision is crucial for coping with the rapid shifts in technology and consumer behavior that characterize modern life.

Moreover, the implementation of intelligent resource allocation has a ripple effect on economic growth. Efficient energy use lowers the overall cost of utilities, making them more affordable for consumers. This increased affordability can, in turn, boost other economic activities as businesses and households have more disposable income to spend elsewhere.

While the potential benefits are immense, the integration of AI in resource allocation also brings challenges. Issues such as data security, privacy, and the ethical use of AI cannot be overlooked. Continuous monitoring and regulatory frameworks are needed to ensure that the technology is used responsibly. Utility companies must also invest in the workforce, providing training and education to staff so they can work effectively alongside these advanced systems.

In summary, AI's role in intelligent resource allocation signifies a monumental shift in how we manage and use critical resources like energy and water. The ripple effects of this technology are far-reaching, impacting efficiencies, costs, environmental sustainability, and even economic growth. While the challenges are non-negligible, the potential for a smarter, greener, and more efficient world makes the endeavor not only worthwhile but imperative for the future.

CHAPTER 24:
ETHICAL CONSIDERATIONS IN AI

As AI continues to permeate our daily lives, the importance of ethical considerations can't be overstated. Concerns around bias and fairness arise when AI systems, often trained on historical data, unintentionally perpetuate or even exacerbate existing inequalities. Privacy concerns are another critical area, as AI systems often require vast amounts of personal data to function effectively, leading to potential misuse or unauthorized access. Moreover, the governance of AI—regulating and guiding its development and deployment—necessitates vigilant oversight to ensure that these technologies benefit society as a whole without compromising individual rights. Navigating these ethical complexities is essential for fostering trust and maximizing the positive impact of AI on everyday life.

Bias and Fairness

In the realm of artificial intelligence, the concepts of bias and fairness emerge as critical considerations, shedding light on the profound complexities and ethical dimensions that trail the seemingly neutral algorithms. At its core, AI learns from data, but this data is often a mirror of society, reflecting both the splendors and the biases inherent within it. Thus, despite the promise of impartiality that automation might suggest, AI systems can perpetuate or even exacerbate social inequalities if not developed with conscientious oversight. The ethical imperative to address bias and champion fairness in AI is not just a technical challenge; it's a fundamental aspect of deploying technology that aligns with our collective values and aspirations for a just society.

Consider the alluring efficiencies introduced by AI in various sectors—from personalized recommendations in digital marketplaces to vital decision-making in criminal justice or hiring. Yet, these intelligent systems run the risk of acting upon biased information, inadvertently becoming vehicles for discrimination. For instance, an AI model trained on employment data with historical gender imbalances may undervalue candidates from underrepresented groups. The crux of the challenge lies in the data: incomplete, non-representative, or historically biased datasets can inadvertently teach the AI erroneous patterns of exclusion or preference.

To bridge the chasm between the current state of AI systems and the ideal of fostering egalitarian tech advancements, multiple methodologies are emerging. One such avenue is the rigorous auditing of AI algorithms, where the system's decisions are meticulously evaluated for fairness and neutrality. Another incipient field tackling this issue is 'fair-ML,' wherein new mathematical models are formulated to tease out and correct biases present in the training data before these biases become entrenched in the AI's decision patterns.

The significance of fairness in AI transcends the operational. It reflects the shared responsibility of developers, users, and policymakers to cultivate digital environments that respect the dignity of all individuals. The challenge is not inconsiderable; human bias is a multi-faceted phenomenon, not easily packaged into neat quantitative parcels. That's why measures to ascertain AI fairness must be continuous and evolving, sensitive to the nuances of cultural contexts, and informed by a diverse spectrum of voices.

Moreover, the call for fairness in AI sparks a bevy of questions: What constitutes fairness in a given context? Who decides the metrics upon which AI systems should be judged? These are not questions with one-size-fits-all answers, but they centralize the fact that discourse on AI fairness must be dialogic and inclusive, drawing upon ethical frameworks and stakeholder input to configure systems equitably.

Yet, fairness is more than an abstract ethical commandment; it is a technical aim, one that requires ground-breaking tools and metrics for its realization. Some such tools might include disparity impact analyses, which scrutinize the differential impacts of AI across varied demographic groups, or accountability mechanisms that trace decisions back to their algorithmic origins. Transparency in how AI systems reach their conclusions is also paramount, as inscrutable 'black box' algorithms can thwart efforts to vet for fairness and bias.

Despite these challenges, the motivations for augmenting AI with fairness are profound. Beyond adhering to legal standards and social expectations, fair AI engenders trust. Trust of consumers in digital marketplaces, trust of citizens in automated public services, and trust of employees in HR algorithms—all hinge upon the assurance that AI systems operate impartially. Consider the immense trust we must place in autonomous vehicles that decide in fractions of a second or health care AI that can sway life-or-death outcomes; such trust necessitates unimpeachable standards of fairness.

Furthermore, fairness in AI holds the promise of leveling the playing field. When properly calibrated, AI can function as an equalizer, offering evidence-based evaluations that may counteract human preconceptions. For instance, anonymizing features in hiring algorithms could foreground merit and abilities over demographic details, or if implemented with care, AI can heighten awareness of unconscious biases, turning a diagnostic lens upon our assumptions and potentially rewiring societal prejudices.

Organizations must also grapple with the operational implementation of fairness. It's not merely about tweaking algorithms; it's about cultivating a fairness-centric culture that informs the AI development lifecycle—from ideation to deployment. Training programs on algorithmic biases, diversity in the workforce building these systems, and ethical guidelines etched into project briefs can all serve as bulwarks against unwitting prejudice.

The pursuit of fairness in AI is a multifaceted and dynamic endeavor, nestled at the confluence of technology, ethics, and sociology. It calls for a coalition of experts—from data scientists to social justice advocates—each providing critical insights into crafting AI that enhances our lives without infringing on our values. It is an ongoing journey, charting a course through uncharted digital territories, with the goal of harnessing the power of AI while steadfastly guarding against its potential to do harm through bias or negligence.

In conclusion, the dialogue on bias and fairness in AI is intrinsic to the broader agenda of ethical AI deployment. It's a testament to the human capacity for using technology not merely as an instrument of change but also as a catalyst for greater equity. As AI becomes increasingly woven into the tapestry of our daily lives, ensuring that it reflects the highest standards of fairness is not just an option; it's an obligation—and an opportunity to reaffighirm our commitment to building a world where technology serves us all, irrespective of our backgrounds or social standing.

Privacy Concerns

Artificial Intelligence (AI) has become an integral part of our daily lives, shaping various facets from healthcare to education, and personal finance to entertainment. While the benefits it offers are undeniable, there's a less explored terrain that demands our attention—privacy concerns associated with the widespread use of AI. As we delve into the ways AI can make our lives easier, it's crucial to address the ever-growing apprehensions about how our data is collected, stored, and used. This chapter will shed light on the essential privacy issues that come with the proliferation of AI technologies, advocating for responsible use and proper safeguards.

First and foremost, the collection of vast amounts of personal data is fundamental to the functioning of AI systems. These technologies thrive on data, and without it, many of the personalized services we appreciate would be impossible. For example, smart assistants like Amazon's Alexa, Apple's Siri, and Google Assistant rely on continuous data input to become more efficient and useful. However, the more data they gather, the more significant the privacy risks. The question then arises: Where do we draw the line between convenience and privacy invasion?

Another vital consideration is the manner in which data is stored and who has access to it. AI systems often employ cloud-based storage solutions, meaning our personal information is stored on servers owned and operated by third-party companies. This raises concerns about data breaches and unauthorized access. Even the most robust security protocols can be vulnerable to sophisticated cyber-attacks, putting personal data at risk. The repercussions of such breaches can be devastating, affecting individuals' personal and financial well-being.

Additionally, there's the issue of consent. In many cases, users might not be fully aware of the extent to which their data is being collected and used. The fine print in user agreements is often overlooked, leaving individuals with little control over their personal information. For instance, AI-driven health monitoring devices gather a plethora of sensitive health data, but users might not always be aware of who has access to this information and how it's being used. This lack of transparency necessitates a call for more straightforward and comprehensible user agreements.

Moreover, AI's ability to analyze and interpret massive datasets can lead to unintended consequences. One area of concern is the inadvertent creation of detailed user profiles that can predict behavior, preferences, and even life choices. While these profiles can enhance user experience by offering personalized recommendations, they can also be intrusive. The risk is that these profiles could be used in ways that infringe on personal privacy or even result in unwanted surveillance. Do we really want every aspect of our lives to be predictable and monitored?

The intersection of AI and privacy extends beyond just data collection and storage. Another critical factor is the regulation and governance of these technologies. Currently, there is a patchwork of regulations that vary significantly between countries and even states. This inconsistency makes it challenging to ensure uniform privacy protections globally. Clear, consistent, and enforceable regulations are necessary to safeguard privacy while allowing innovation to flourish. Policymakers, technologists, and ethicists must work together to design frameworks that balance these competing needs.

Ethical use of AI also demands that companies prioritize user privacy throughout the development and deployment phases of AI applications. This involves adopting practices such as data minimization, where only the data necessary for the application's functionality is collected. Developers should also implement strong encryption methods to protect data and ensure that user data is anonymized wherever possible. Ethical AI design should prioritize user consent and control, giving individuals the ability to easily manage and understand their data footprint.

Another dimension to consider is the potential misuse of AI by malicious entities. Cybercriminals, for instance, could exploit AI systems to gather sensitive data for nefarious purposes. Sophisticated AI algorithms can be used to conduct phishing attacks, identity theft, and other cybercrimes. As AI capabilities continue to advance, so too do the tactics of those looking to exploit these technologies for harm. This evolving threat landscape makes it all the more critical to invest in robust security measures and continuous monitoring of AI systems.

In the realm of AI and privacy, another emerging concern is facial recognition technology. Used in everything from law enforcement to social media tagging, facial recognition systems gather and process biometric data, which is incredibly personal. The use of such technology raises significant privacy issues, especially when deployed without explicit consent or for surveillance purposes. The potential for abuse in this area is vast, underscoring the need for stringent guidelines and regulations.

Furthermore, AI's impact on privacy isn't confined to individuals alone; it also affects groups and communities. For instance, predictive policing algorithms that rely on historical crime data can disproportionately target specific communities, leading to privacy infringements and a sense of being constantly monitored. Similarly, social media platforms that use AI for content personalization might inadvertently contribute to echo chambers, where users' data is used to feed them biased information, thereby infringing on their intellectual privacy.

To strike a balance between leveraging AI's advantages and protecting privacy, there is a need for widespread digital literacy. Educating the public about how AI works, what data it collects, and how this data is used can empower individuals to make informed decisions. Digital literacy initiatives should be aimed at demystifying AI and equipping people with the knowledge to safeguard their privacy effectively.

Lastly, fostering a culture of ethical AI development is paramount. Companies and developers should be encouraged—and in some cases mandated—to integrate privacy considerations into every stage of AI development. This 'privacy by design' approach should be a standard practice, ensuring that new AI technologies are built with privacy protection in mind from the outset.

The promise of AI is immense, but it brings with it significant privacy challenges that cannot be ignored. By fostering open dialogue, implementing robust regulations, and prioritizing ethical development, we can harness the power of AI while safeguarding our privacy. This delicate balance is essential for the sustainable and ethical advancement of AI technologies in our everyday lives.

AI Governance

AI Governance stands as a crucial pillar in the broader discussion of ethical considerations in AI. As AI systems become more sophisticated and embedded in various aspects of everyday life, the need for robust governance structures becomes increasingly urgent. AI governance refers to the frameworks, policies, and regulations designed to oversee the development, deployment, and use of AI technologies, ensuring they serve the public good while minimizing risks.

Establishing comprehensive AI governance is no small feat. It involves a multidisciplinary approach that encompasses legal frameworks, ethical guidelines, technical standards, and societal norms. This multifaceted strategy aims to strike a balance between innovation and regulation, allowing technological advancements while safeguarding human rights, privacy, and fairness.

One vital component of AI governance is the establishment of ethical guidelines. Ethical standards provide a moral compass for AI developers, helping them navigate the complex landscape of AI design and deployment. These guidelines typically address issues such as transparency, accountability, and bias. Transparent AI systems make it easier for users to understand how decisions are made, fostering trust between humans and machines.

Accountability in AI governance ensures that there is always a human or organizational entity responsible for the actions and decisions made by AI systems. This is particularly important in high-stakes areas like healthcare, finance, and criminal justice, where the impact of AI decisions can be profound. Ensuring accountability helps in the effective management of risks associated with AI.

Bias and fairness are critical challenges addressed by AI governance frameworks. AI systems are trained on data, and if this data is biased, the system's outcomes will likely reflect those biases. Governance structures must therefore include mechanisms for identifying, mitigating, and correcting biases in AI systems. This is essential for promoting fairness and preventing discrimination in AI applications.

Privacy concerns are another significant aspect of AI governance. With the increasing capabilities of AI systems to analyze vast amounts of personal data, protecting individual privacy has never been more crucial. Regulatory frameworks, such as the General Data Protection Regulation (GDPR) in the European Union, are examples of how laws can be tailored to safeguard personal data and ensure that AI systems are used responsibly.

Technical standards and best practices are integral to effective AI governance. These standards guide the development and deployment of AI systems, ensuring they operate reliably and safely. Industry bodies and international organizations play a crucial role in developing these standards, facilitating global cooperation and harmonization in AI governance.

Another key element of AI governance is public engagement and education. Democratically governed AI requires the active participation of the public in shaping the policies and guidelines that govern AI technologies. Public consultation processes and educational campaigns are essential in informing citizens about AI developments and involving them in meaningful discussions about its implications.

Global cooperation is also imperative for coherent and effective AI governance. AI technologies do not adhere to national boundaries, making international collaboration necessary. Multilateral initiatives, such as those led by the United Nations and the Organisation for Economic Co-operation and Development (OECD), are examples of efforts to create universal standards and promote responsible AI use worldwide.

AI governance also addresses the need for regulatory bodies and oversight institutions. These entities are responsible for monitoring AI developments, enforcing regulations, and ensuring compliance. They serve as watchdogs, ensuring that AI systems are deployed ethically and transparently while holding developers and organizations accountable for their actions.

The evolving nature of AI technology necessitates adaptive and forward-thinking governance frameworks. Regulations must not stifle innovation but rather create an environment where ethical AI can flourish. This involves a delicate balancing act of updating policies to keep pace with technological advancements while providing room for experimentation and growth.

For AI governance to be truly effective, it requires a collaborative effort from all stakeholders, including governments, private sector companies, academic institutions, and civil society. Each of these actors brings valuable perspectives and expertise to the table, contributing to a holistic approach to governing AI.

In conclusion, AI governance is a foundational aspect of building a future where AI technologies are aligned with human values and societal goals. By establishing robust ethical guidelines, ensuring accountability, addressing bias and privacy concerns, adopting technical standards, promoting public engagement, fostering global cooperation, and creating effective oversight mechanisms, we can harness the transformative potential of AI while mitigating its risks. The journey towards comprehensive AI governance is ongoing, but it is a necessary endeavor to ensure that AI contributes positively to society and enhances our everyday lives.

CHAPTER 25:
THE FUTURE OF AI IN EVERYDAY LIFE

As we look ahead, the future of AI in everyday life promises to be both revolutionary and intricate, intertwining seamlessly with the routines and activities that define our daily existence. Emerging trends like AI-enhanced personal assistants and advanced smart home systems suggest a world where AI anticipates and addresses our needs before we're even aware of them. However, with these advancements come challenges and opportunities that demand our attention—issues of privacy, ethical governance, and digital literacy must be tackled head-on. Being well-prepared for an AI-driven future means embracing the technology with an open, yet discerning mindset, ensuring that we harness its potential for the betterment of society while safeguarding our fundamental values and rights. This balanced approach will allow AI to not only enhance our efficiency and convenience but also contribute to a more informed and connected world.

Emerging Trends

The evolution of AI has been nothing short of remarkable, and it shows no signs of stopping. As we look forward, several emerging trends in AI hold the potential to vastly transform everyday life, making it more efficient, personalized, and interconnected.

One of the most ground-breaking trends is the rise of AI-driven edge computing. Unlike traditional cloud computing, where data is processed in centralized data centers, edge computing brings the pro-

AI Everyday

cessing power closer to the source of data generation. This decentrali-
zation is particularly relevant for smart homes and IoT (Internet of
Things) devices. Imagine your smart thermostat or fitness tracker being able to make split-second decisions without any lag, improving responsiveness and efficiency. Edge computing can potentially reduce
latency and bandwidth usage, leading to quicker and more reliable AI
applications right in our homes.

Another trend to keep an eye on is the increasing sophistication of
natural language processing (NLP). While voice assistants like Siri and
Alexa have gotten much better at understanding us, the future promises even more advancements. Think about being able to hold a conversation with your smart assistant that's nearly indistinguishable from
chatting with a human. This level of sophistication can revolutionize
customer service, emotional support applications, and even personal
companionship in the form of AI friends or assistants who understand
not just the words, but the context and emotions behind them.

The growth of personalized AI is another noteworthy trend. From
custom-tailored workouts to financial advice tailored to your spending
habits, AI is getting better at figuring out what you need before you
even realize it. In healthcare, for instance, algorithms are being designed to offer highly personalized medical treatments based on an individual's genetic makeup, lifestyle, and medical history. Personalized
AI can improve outcomes and efficiency in virtually every aspect of
life, making services more relevant and beneficial.

Privacy and data security continue to be major areas of focus. As
AI becomes more integrated into our lives, the need for secure,
privacy-respecting technologies becomes crucial. Emerging techniques
like federated learning allow for the development of AI models using
data from multiple sources without actually collecting and sharing that
data. This approach enhances privacy and reduces the risk of data

breaches while still enabling the continuous improvement of AI systems.

In the world of education, AI is making strides in creating adaptive learning environments. These systems can dynamically adjust the difficulty of content based on the learner's progress, ensuring a more personalized and efficient learning experience. This can be especially beneficial in remote or underfunded educational scenarios, providing high-quality, adapted educational resources to everyone.

Moreover, AI is increasingly blending with augmented reality (AR) and virtual reality (VR), creating immersive experiences that were once the domain of science fiction. Whether it's a virtual personal trainer guiding you through a workout or an AR app that provides real-time language translation, these technologies are set to make our interactions more interactive and engaging. Imagine touring a virtual rendition of your future home or having an AI-driven coach who monitors your form and progress in real-time, all within a completely immersive environment.

The corporate landscape is also being reshaped by AI, with intelligent automation and robotic process automation (RPA) playing central roles. These technologies are streamlining workflows by automating repetitive, mundane tasks, allowing employees to focus on more creative and strategic endeavors. This not only boosts productivity but also enhances job satisfaction and drives innovation.

Blockchain technology and AI are increasingly converging, leading to advancements that could redefine data integrity, security, and transparency. Decentralized AI systems harness the power of blockchain to create more secure and trustworthy environments. Whether it's for secure health records management or transparent supply chains, this amalgamation offers groundbreaking solutions that are resilient against fraudulent activities.

Environmental sustainability is another critical area where AI is making impactful strides. From smarter energy grids that optimize power distribution to AI-driven systems that monitor and reduce waste, the potential for AI to contribute towards a greener planet is immense. Emerging technologies are continuously finding new ways to minimize human impact on the environment, driving us toward a more sustainable future.

Telehealth and remote medical consultations, accelerated by the COVID-19 pandemic, are expected to remain a vital aspect of healthcare, thanks to AI. With improved diagnostics and patient monitoring via AI, healthcare professionals can make more accurate decisions even from a distance. This trend ensures that access to quality healthcare isn't limited by geography, enabling rural areas and underserved communities to receive the medical attention they need.

The democratisation of AI tools is another trend making waves. Open-source AI platforms and easier access to AI development resources mean that almost anyone with a great idea can contribute to the AI revolution. This influx of diverse applications developed by various individuals and smaller organizations democratizes innovation, ensuring that advancements are not restricted to big tech companies but are accessible to a broader population.

5G networks are set to bolster the capabilities of connected AI systems. The increased speed and reduced latency offered by 5G will enhance the performance of smart devices, making real-time AI applications a reality. From seamless video calling and live translations to instantaneous cloud gaming, 5G will enable a new wave of AI-enhanced experiences that were previously unimaginable.

AI ethics is emerging as a crucial field of study and implementtation. There is an increasing awareness of the importance of developing ethical guidelines and frameworks to ensure that AI technologies are designed and deployed responsibly. This ensures that as AI permeates more aspects of our lives, its impact is positive and equitable. Topics like bias reduction, transparency, and accountability are gaining traction as essential elements in the responsible deployment of AI technologies.

In sum, the future of AI in everyday life is thrilling and multifaceted, promising enhancements across many aspects of our existence. While challenges certainly lie ahead, the emerging trends in AI point to a future where technology and human experience are seamlessly intertwined, paving the way for a smarter, more efficient, and compassionate world.

Challenges and Opportunities

As we move further into an era where artificial intelligence becomes an integral part of our daily lives, we're faced with an array of challenges and opportunities. Navigating these waters requires a balanced understanding of both the potential and the pitfalls of AI. For every door that AI opens, there are hurdles that cannot be ignored.

One of the significant challenges is the ethical implications surrounding AI. We must ensure fairness and impartiality in AI algorithms. When AI decisions impact real-life scenarios - from job recruitment to loan approvals - ensuring these processes are unbiased is crucial. Addressing biases in AI isn't merely a technical issue; it is a societal one that demands collective accountability and scrutiny.

While ethical considerations present one set of challenges, data security is another major concern. AI systems rely on vast amounts of data for training, which often includes sensitive personal information. Protecting this data from breaches and ensuring user privacy is imperative. As AI becomes more widespread, regulations and standards need to evolve to safeguard against misuse and ensure responsible usage.

On the flip side, the capability of AI to revolutionize industries offers unparalleled opportunities. In healthcare, for instance, AI can lead to breakthroughs in personalized medicine, enabling treatments tailored to individual genetic makeups. This possibility holds the promise of more effective and efficient healthcare solutions, reducing costs and improving outcomes.

Similarly, in education, AI's potential to create adaptive learning technologies means that each student's educational experience can be customized to their unique needs, pace, and learning style. This individualized approach can lead to a more engaging and effective learning environment, enhancing educational outcomes and closing achievement gaps.

Another area ripe with opportunities is AI in environmental sustainability. AI-powered systems can analyze vast datasets to optimize energy usage, manage waste more efficiently, and monitor environmental changes in real-time. Such advancements could play a pivotal role in tackling climate change and promoting sustainable practices across industries.

The workforce is another domain where AI presents both daunting challenges and exciting opportunities. Automation and AI-driven systems could displace certain job categories, leading to economic and social disruptions. However, they also open up new avenues for job creation in emerging fields, necessitating a shift in how we approach education and workforce training.

In transportation, autonomous vehicles and intelligent traffic management systems promise to enhance safety, reduce congestion, and lower environmental impact. These advancements could transform urban life, making cities smarter and more efficient. However, they also present regulatory challenges and require robust frameworks to ensure safety and reliability.

AI's potential to enhance productivity in various sectors can't be overstated. From intelligent task management systems to virtual assistants, AI tools can streamline workflows, reduce human error, and increase efficiency. The challenge lies in ensuring that these tools complement human workers rather than replace them, fostering a collaborative human-AI working environment.

Retail and shopping also stand to benefit significantly from AI. Personalized shopping experiences driven by AI can improve customer satisfaction and loyalty. AI's role in supply chain management can optimize inventory levels and predict consumer demand more accurately. However, integrating such systems requires overcoming technical, logistical, and sometimes even cultural barriers within organizations.

While the opportunities are immense, fostering a responsible AI-driven future is pivotal. This involves not only technological innovation but also developing policies, regulations, and educational programs that keep pace with AI advancements. Preparing for this AI-driven future is a collective effort requiring the participation of governments, industries, and individuals alike.

One thrilling aspect of AI is its potential in mental health and wellness. With AI-driven tools for mental health monitoring and personalized therapy plans, we can address mental health challenges more proactively and effectively. However, the delicate nature of mental health data requires stringent privacy protections and ethical considerations.

The integration of AI in personal finance is another promising area. Automated budgeting tools and AI-driven investment platforms can help individuals make informed financial decisions. Fraud detection systems can add an extra layer of security, protecting our finances. Nevertheless, these technologies must ensure transparency and be user-friendly to gain widespread adoption.

AI in smart cities represents a blend of opportunities and challenges. Intelligent systems can enhance public safety, improve city services, and optimize resources. However, achieving this requires addressing technological, infrastructural, and social challenges. Implementing such systems at scale demands robust infrastructural investments and public trust in AI technologies.

Entertainment is yet another area where AI can provide immense value. AI-driven streaming recommendations and personalized content can significantly enhance user experiences. In gaming, AI can create more immersive and challenging experiences. However, balancing personalization with user privacy and avoiding over-reliance on algorithmic curation remains a challenge.

The travel and tourism industry can leverage AI for personalized travel plans, customer service, and smart travel apps. These innovations can make travel experiences smoother and more enjoyable. However, ensuring data security and maintaining a human touch in customer interactions are essential to avoid a purely transactional approach.

AI in agriculture holds promise for addressing food security and sustainability challenges. From precision farming to automated equipment, AI can significantly boost agricultural productivity and resource efficiency. However, addressing the technological divide between large and small-scale farmers is crucial to ensure equitable access to AI benefits.

In summary, AI presents a landscape rich with opportunities but not without its challenges. Ethical considerations, data security, workforce impacts, and regulatory frameworks are critical issues that need addressing. At the same time, the potential for AI to transform healthcare, education, environmental sustainability, and various other sectors offers a glimpse into an exciting future where AI enhances everyday life. Embracing this future requires a multi-faceted approach that balances innovation with responsibility, ensuring that the benefits of AI are accessible, equitable, and sustainable.

Preparing for an AI-Driven Future

The wave of AI technology is transforming our daily lives in ways we could only have imagined a few years ago. With AI becoming an integral part of home automation, healthcare, education, personal finance, and even entertainment, an important question arises: how do we prepare for this AI-driven future?

First and foremost, understanding AI at a fundamental level is a crucial step. There's a difference between using AI systems and truly comprehending their potential and limitations. Many people use virtual assistants like Alexa or Siri without considering the complex algorithms and data processing behind these tools. Developing a basic knowledge of AI concepts and terminologies can demystify these technologies, making them less intimidating and more accessible.

Adapting to an AI-driven world doesn't mean we all need to become data scientists or engineers. However, being proactive in learning and embracing AI technologies will position us better for future advancements. Online courses, webinars, and seminars are readily available for those who are eager to dive deeper into AI's intricacies. Even casual learners can find simplified, user-friendly resources to get started.

One practical way to prepare is by gradually integrating AI tools into your everyday activities. Start with areas where AI applications are most prevalent and have proven benefits. For instance, utilize smart home devices to automate mundane tasks, making your living space more efficient and enjoyable. Try out AI-driven finance tools to streamline budgeting and investments, enhancing your financial literacy and control.

Healthcare is another domain where AI offers significant advantages. Wearable health monitoring devices and personalized medicine are just the beginning. By using these tools, you not only enjoy better health management but also familiarize yourself with the operational aspects of AI in healthcare. Simply interacting with these technologies can build your comfort and competence over time.

Preparation also involves staying updated with emerging trends and technologies. AI is continuously evolving, and what might be cutting-edge today could become obsolete tomorrow. Keeping abreast of the latest advancements can help you anticipate changes and adjust accordingly. Subscribing to tech journals, following AI research news, and participating in tech communities can keep you informed and in the loop.

Embracing an AI-driven future also calls for a mindset shift. We need to overcome the fear of AI replacing human jobs and instead focus on how these technologies can complement and enhance our abilities. AI isn't just about automation; it's about augmentation—enhancing human capabilities and enabling us to achieve tasks we couldn't before. Developing a positive and adaptive mindset can greatly facilitate this transition.

On the professional front, acquiring skills that are complementary to AI is a strategic move. Skills in data analysis, machine learning, and even basic programming can significantly enhance your employment prospects. Employers today value workers who can understand and leverage AI to optimize workflows and drive innovation.

It's also crucial to cultivate soft skills like critical thinking, emotional intelligence, and creativity. AI excels at tasks that are repetitive, data-intensive, and predictable. However, human intuition, empathy, and inventive thinking remain irreplaceable. These unique human attributes will become even more valuable as AI takes over routine tasks, allowing us to focus on higher-order activities.

Another aspect of preparation involves public policy and ethical considerations. It's imperative to engage in discussions about the ethical use of AI, data privacy, and the potential biases in AI algorithms. Being an informed citizen means participating in these conversations and advocating for responsible AI governance. Policies and regulations will shape the trajectory of AI development, and everyone's voice matters in this dialogue.

Finally, let's not underestimate the societal implications of AI. The impact on various sectors from healthcare to transportation, and even entertainment, is profound. By preparing ourselves individually, we contribute to a collective readiness. Communities, industries, and nations need to work together to ensure that the benefits of AI are equitably distributed, and the challenges addressed responsibly.

In essence, preparing for an AI-driven future is about fostering an environment of learning, adaptability, and responsible use of technology. The journey involves an ongoing commitment to understanding and skill development, balanced with ethical and societal considerations. With the right approach, we can harness AI's transformative power to improve and simplify our everyday lives, creating a future where technology and humanity thrive together.

Conclusion

As we reach the conclusion of our exploration into the transformative potential of Artificial Intelligence, it's evident that AI is more than just a collection of futuristic concepts and technological buzzwords. It's a rapidly evolving tool with the capability to dramatically enhance various aspects of our everyday lives. Whether it's simplifying daily tasks, improving our healthcare, or even revolutionizing the way we communicate, AI's practical applications are boundless and profound.

In our homes, AI-driven smart assistants and automation systems are creating environments that respond intuitively to our needs. These technologies are no longer the stuff of science fiction; they're real and improving at an unprecedented pace. The same can be said for how AI is reshaping healthcare, making personalized medicine and advanced diagnostics more accessible than ever before. By integrating AI into our medical systems, we not only enhance patient care but also open new frontiers in addressing complex health challenges.

Education is another realm where AI is making waves. Adaptive learning technologies and virtual classrooms offer customized educational experiences that cater to individual learning styles and paces. This personalized approach is revolutionizing how knowledge is imparted and absorbed, making education more inclusive and effective. AI tutors, for instance, provide students with real-time feedback and customized content, bridging gaps that traditional teaching methods might overlook.

When it comes to managing personal finances, AI offers tools that bring efficiency and intelligence to budgeting and investment decisions. Automated budgeting tools help individuals manage their expenses effortlessly, while AI-driven investment platforms provide sophisticated analyses that were once only available to institutional investors. Fraud detection algorithms enhance the security of our financial transactions, giving us peace of mind in an increasingly digital world.

In the realm of entertainment, AI enhances our experiences by curating content to match our tastes and preferences. Streaming services use complex algorithms to suggest movies and shows we'll likely enjoy, while video game developers employ AI to create more immersive and adaptive gaming experiences. Even virtual reality is getting a boost from AI, making these experiences more realistic and engaging.

As we commute and travel, AI-driven advancements in transportation are making our journeys safer and more efficient. Autonomous vehicles, intelligent traffic management systems, and optimized ride-sharing services are not just on the horizon—they're already here, fundamentally changing how we move from place to place. These technologies promise to reduce traffic congestion, lower emissions, and improve overall transportation efficiency.

In retail and shopping, AI creates personalized experiences that cater to our preferences, making shopping more enjoyable and efficient. It optimizes supply chain processes and manages inventories smartly, ensuring that products are available when and where we need them. Smart warehouses minimize human error and maximize operational efficiency, revolutionizing the backend of retail logistics.

Work and productivity are also benefiting from AI's relentless march. Intelligent task management tools and virtual assistants streamline our workflows, allowing us to focus on more creative and strategic tasks. Whether it's project management or day-to-day operations, AI tools enhance productivity and help maintain work-life balance by handling mundane tasks more efficiently.

Even in social media, AI plays a critical role in personalizing content, analyzing data, and automating interactions. These capabilities allow for more meaningful online engagements and provide insights that empower users and businesses alike. AI-driven analytics offer deep dives into user behavior, enabling more strategic decision-making and targeted communication.

The influence of AI extends into real estate, travel, fitness, agriculture, environmental sustainability, government services, legal services, marketing, human resources, manufacturing, logistics, and energy and utilities. Each sector sees unique benefits from AI's implementation, ranging from smart property management and precise travel planning to optimized energy forecasts and intelligent resource allocations. The potential is vast and the implications are profound, heralding a future where AI is integral to progress and innovation.

As we look toward the future, it's crucial to address the ethical considerations accompanying AI's proliferation. Bias, fairness, privacy, and governance are pertinent issues that demand our attention and action. Establishing ethical frameworks and balanced regulations will ensure that AI develops in a manner that benefits society as a whole while mitigating potential risks.

In conclusion, the journey through AI's landscape reveals a technology that's not merely augmenting our life but fundamentally enhancing it, paving the way for a future where AI's role is integral and indispensable. The promise of AI lies not just in its ability to perform tasks but in its potential to create a world that's more efficient, accessible, and enriching for everyone. Embracing these advancements, while consciously navigating the ethical terrain, sets the stage for a future where AI doesn't just support us but elevates the very essence of our daily living.

APPENDIX A:
APPENDIX

The appendix serves as a valuable catch-all section, encapsulating critical addendums to the main content of this book. Here, you'll find resources that bring additional clarity and support to the topics discussed throughout, providing a deeper understanding of how Artificial Intelligence (AI) can be leveraged in everyday life. This section contains supplementary materials such as detailed charts, references, frameworks, and checklists designed to enhance your grasp of AI's multifaceted applications.

Additional Resources

We've compiled a list of recommended readings, online courses, and other educational resources to further your knowledge of AI. Whether you're a beginner looking to understand the basics or an advanced learner interested in specialized areas, these resources offer something for everyone.

Books and Articles: A curated selection of insightful reads that dive deeper into various AI topics covered in the book.

Online Courses: Educational platforms offering courses ranging from introductory to advanced levels in AI and machine learning.

Webinars and Lectures: Informative sessions by experts in the field, discussing current trends and future directions of AI.

Tools and Frameworks

Utilizing the right tools and frameworks can significantly accelerate your journey in mastering AI. This section lists out commonly used AI tools and frameworks, along with brief descriptions and links for further exploration.

TensorFlow: An open-source platform for machine learning, ideal for building and training models.

PyTorch: A flexible deep learning framework that allows for building complex architectures with ease.

Scikit-learn: A simple and efficient tool for data mining and data analysis, built on NumPy, SciPy, and Matplotlib.

Glossary of Terms

AI comes with its own set of terminologies and jargons. We've put together a comprehensive glossary of terms to help you navigate through the technical language used throughout the book. This glossary aims to demystify complex terms and make the content accessible to all readers.

Case Studies

Real-world examples provide valuable insights into how AI is being implemented across different industries. In this section, we present case studies that illustrate the transformative power of AI in various applications, from healthcare and finance to entertainment and agriculture.

Case Study 1: AI in Healthcare - Personalized Treatment Plans

Case Study 2: AI in Finance - Fraud Detection Systems

Case Study 3: AI in Agriculture - Precision Farming Techniques

Checklists

We've created checklists to help you implement AI solutions in your own life or organization. These practical guides focus on key steps and considerations for successful AI adoption and integration.

Steps to Implement AI in Business

Checklist for Home Automation with AI

Considerations for Ethical AI Deployment

By delving into the materials provided in this appendix, you'll equip yourself with the additional knowledge and tools needed to harness the full potential of AI in everyday life. The appendices complement the book's main chapters, rounding out your learning experience and empowering you to take meaningful steps toward an AI-enhanced future.

GLOSSARY OF TERMS

This glossary aims to clarify common terms and concepts related to Artificial Intelligence (AI). Use it as a reference to enhance your understanding as you explore the various applications of AI in daily life.

AI (Artificial Intelligence): The simulation of human intelligence processes by machines, especially computer systems. These processes include learning, reasoning, and self-correction.

Adaptive Learning: Educational systems that use AI to tailor learning experiences to the individual needs of each student, often by modifying content in real-time.

Algorithm: A set of rules or steps to be followed in problem-solving or calculation, often implemented by computer software.

Autonomous Vehicle: A car or other vehicle capable of navigating and driving itself without human intervention, primarily using AI and sensor technology.

Bias (AI Bias): Systemic and repeatable errors in a computer system that create unfair outcomes, such as privileging one arbitrary group over others.

Big Data: Large and complex data sets that traditional data-processing software cannot deal with. AI can analyze these data sets to find patterns and insights.

Chatbot: An AI program designed to simulate conversation with human users, especially over the internet. Often used for customer service.

Deep Learning: A subset of machine learning in AI that features networks capable of learning unsupervised from unstructured or unlabeled data.

Machine Learning: A type of AI that enables a system to learn from data rather than through explicit programming. Focuses on the development of computer programs that can access data and use it to learn for themselves.

Natural Language Processing (NLP): A field of AI that enables computers to understand, interpret, and respond to human languages in a valuable way. It's used in applications like chatbots and language translation services.

Neural Network: A computer system modeled on the human brain and nervous system. It's designed to recognize patterns and solve problems in a manner similar to the human brain.

Predictive Analytics: Utilizes statistical algorithms and AI techniques to identify the likelihood of future outcomes based on historical data. Often used in finance, marketing, and healthcare.

Robotics: The design, construction, operation, and use of robots. AI is integral in enabling robots to perform complex tasks autonomously.

Smart Assistant: An AI program that can perform tasks or services for an individual based on commands or questions. Examples include Amazon's Alexa and Apple's Siri.

Supervised Learning: A type of machine learning where the model is trained on labeled data. The system is provided with input-output pairs and learns to map inputs to the correct outputs.

Unsupervised Learning: A type of machine learning where the model is given unlabeled data and must find patterns and relationships within it without external guidance.

Consult this glossary as you delve into the expansive world of AI applications, gaining a better grasp of the terminology that shapes this transformative technology.

Resources for Further Reading

Delving deeper into the realm of Artificial Intelligence can significantly broaden your understanding and appreciation of its transformative power. Although this book aims to provide a comprehensive introduction to various applications of AI, there are numerous other valuable resources that can offer deeper insights into specific areas. These range from foundational texts and cutting-edge research papers to online courses and expert blogs. Each has its unique contribution to your learning journey.

Books remain one of the foundational resources for acquiring in-depth knowledge. Experts like Stuart Russell and Peter Norvig authored "Artificial Intelligence: A Modern Approach," a staple for anyone keen on the technical facets of AI. On the more philosophical side of things, Nick Bostrom's "Superintelligence: Paths, Dangers, Strategies" offers potent insights into long-term implications and ethical considerations around AI. For those interested in real-world applications and economic impacts, Kai-Fu Lee's "AI Superpowers: China, Silicon Valley, and the New World Order" is incredibly illuminating.

For a blend of approachable and illuminating content, various online platforms offer free and paid courses taught by industry experts. Platforms like Coursera, edX, and Udacity partner with top universities and industry leaders to provide specialized courses. From the basics of machine learning algorithms to sophisticated applications in natural language processing, these courses often use a mix of video lectures, interactive quizzes, and hands-on projects designed to solidify your understanding.

Research papers are another invaluable resource, especially for those with a more academic focus. Websites like arXiv.org offer open-access papers on a wide range of AI topics, from theoretical advancements to practical applications. Google Scholar can also serve as a go-to search engine for academic papers, helping you stay up-to-date with the latest scientific discoveries and technological breakthroughs.

Following influential blogs and industry publications is another effective way to continue learning. Websites like Towards Data Science, Medium, and AI News offer articles written by both industry veterans and emerging experts. They cover a wide range of topics, from coding tutorials to opinion pieces on the ethical implications of AI. These resources are often updated frequently, providing fresh perspectives and the latest developments in the field.

Participating in online forums and communities can also be instrumental for your growth. Websites such as Stack Overflow, Reddit, and specialized forums like AI Alignment Forum or ResearchGate enable you to ask questions, share knowledge, and engage in discussions with peers and experts worldwide. This can be particularly beneficial for troubleshooting issues, gaining insights into best practices, and networking with other AI enthusiasts.

Listening to podcasts is a convenient way to keep learning, especially if you have a hectic schedule. Shows like "AI Alignment Podcast," "The AI Alignment Podcast," and "This Week in Machine Learning & AI" feature interviews with leading researchers, industry practitioners, and thought leaders. These podcasts cover a broad spectrum of topics, from technical aspects and career advice to ethical considerations and future trends.

Subscribing to newsletters is another excellent method to stay informed. Publications like The Batch by Deeplearning.ai, and O'Reilly AI Newsletter deliver curated content right to your inbox. These newsletters often include a mix of blog posts, research updates, and information about upcoming events, providing a well-rounded view of the AI landscape.

Academic institutions and research organizations also provide a wealth of accessible materials. Universities like Stanford, MIT, and Carnegie Mellon offer open courseware and lecture series that can serve as valuable learning tools. Additionally, research labs such as OpenAI, DeepMind, and Google AI frequently publish their findings, offering a glimpse into cutting-edge research and emerging trends.

Participating in workshops, conferences, and webinars is another practical way to engage with the AI community. Events like NeurIPS, ICCV, and the International Joint Conference on Artificial Intelligence (IJCAI) gather experts from around the world to discuss the latest research and advancements. Webinars hosted by organizations like IEEE and ACM can also provide valuable learning opportunities and allow for real-time interaction with thought leaders.

To expand your practical skills, engaging in coding competitions and hackathons can be particularly beneficial. Platforms like Kaggle host competitions that push you to apply your knowledge to real-world problems while offering a collaborative environment to learn from others. These experiences not only sharpen your problem-solving skills and coding capabilities but also enhance your ability to work under pressure.

Lastly, ethical considerations in AI are growing in importance and warrant dedicated study. Books such as Cathy O'Neil's "Weapons of Math Destruction" and resources from the AI Ethics Lab can offer a balanced perspective on how to develop and deploy AI responsibly. Understanding the societal implications of AI will ensure you can contribute meaningfully to the ongoing discourse around its use and regulation.

Whether you prefer books, online courses, research papers, or interactive communities, the key is to remain curious and proactive in your learning journey. The field of AI is rapidly evolving, and keeping yourself updated with the latest tools, techniques, and discussions will empower you to leverage AI effectively in your daily life and professional endeavors.

In summary, the resources mentioned above are just the tip of the iceberg. As you continue to explore, you'll find an abundance of information tailored to diverse aspects of AI, each contributing to a fuller, richer understanding of this transformative technology. Happy learning!

www.ingramcontent.com/pod-product-compliance
Lightning Source LLC
Chambersburg PA
CBHW051224050326
40689CB00007B/798